W9-CKT-971

Selected Titles in This Series

Volume

Confoliations

University
LECTURE
Series

Volume 13

Confoliations

Yakov M. Eliashberg
William P. Thurston

American Mathematical Society
Providence, Rhode Island

Partially supported by the NSF grant DMS-9626430
and J. S. Guggenheim Memorial Foundation;
partially supported by the NSF grant DMS-9022140.

1991 *Mathematics Subject Classification.* Primary 53C15, 57N10,
Secondary 58F05, 57R30.

ABSTRACT. The theory of contact structures and the theory of foliations have developed rather independently. They come from separate traditions, and have different flavors. However, in dimension 3, they have evolved in parallel directions that have powerful topological applications involving *tight contact structures* on the one hand and *taut foliations* on the other.

The present work develops the foundations for a theory of *confoliations* to link these two theories, with the aim of developing a combined toolkit that includes both the strongly geometric constructions characteristic of foliation theory and the analytic tools, the connections to four-dimensional topology, and the flexibility characteristic of the theory of contact structures.

In particular, we prove that every C^2 taut foliation can be C^0-perturbed to give a tight contact structure.

Library of Congress Cataloging-in-Publication Data

Eliashberg, Y., 1946–
    Confoliations / Yakov M. Eliashberg, William P. Thurston.
        p. cm. — (University lecture series, ISSN 1047-3998 ; v. 13)
    Includes bibliographical references.
    ISBN 0-8218-0776-5
    1. Foliations (Mathematics) 2. Three-manifolds (Topology) I. Thurston, William P.,
1946– . II. Title. III. Series: University lecture series (Providence, R. I.) ; 13.
QA613.62.E45 1997
514'.72—dc21

97-32128
CIP

Contents

Introduction

In these notes we present the first steps of a *theory of confoliations* which is designed to link the geometry and topology of 3-dimensional contact structures with the geometry and topology of codimension 1 foliations on 3-manifolds. A confoliation is a mixed structure which interpolates between a codimension 1 foliation and a contact structure. This object (without the name) first appeared in work of Steve Altschuler (see [2]). In this paper we will be concerned exclusively with the 3-dimensional case, although confoliations can be defined on higher dimensional manifolds as well (see Section 1.1.6 below).

Foliations and contact structures have been studied practically independently. Indeed at the first glance, these objects belong to two different worlds. The theory of foliations is a part of topology and dynamical systems while contact geometry is an odd-dimensional brother of symplectic geometry. However, the two theories have developed a number of striking similarities. In each case, an understanding developed that additional restrictions are important on foliations and contact structures to make them interesting and useful for applications, for otherwise, these structures are so flexible that they fit anywhere, hence produce no information about the topology of the underlying manifold. This extra flexibility is caused by the appearance of *Reeb components* in the case of foliations (see [47], [55], [53]), and by the *overtwisting phenomenon* in the case of contact structures (see [6], [10]). In order to make the structures more rigid in the context of foliations the theory of *taut foliations* was developed ([52], [51] [19]), as well as the related theory of essential laminations ([20], [31]). In the parallel world of contact geometry, the theory of *tight contact structures* were developed for similar purposes ([6], [11], [23]).

The theory of confoliations should help us to better understand links between the two theories and should provide an instrument for transporting the results from one field to the other.

Acknowledgements. The authors benefited a lot from discussions with many specialists in both fields. We are especially grateful to E. Ghys, E. Giroux, A. Hatcher and I. Katznelson. E. Ghys proved Theorem 1.2.7 as an answer to our question. We thank R. Hind for attentive reading of the manuscript and A. Haefliger for many critical comments, useful suggestions and for providing us with several references.

CHAPTER 1

Geometric nature of integrability

1.1. Foliations, contact structures and confoliations

1.1.1. Definitions. Unlike a line field, a two-dimensional plane field ξ on a 3-dimensional manifold M is not necessarily integrable, that is, there are not necessarily surfaces, even locally, whose tangent planes are in ξ.[1] When ξ is integrable, then its integral surfaces form a *foliation* of the manifold M, so that locally it looks as a fibration of the cylinder $D^2 \times I$ by the horizontal discs $D^2 \times t$, $t \in I$. Usually, we will not distinguish between a foliation and the plane field ξ tangent to it, and often call ξ itself a foliation, if it is integrable.

To understand better the geometry of a 2-dimensional plane field and the geometric meaning of integrability, it is helpful to consider the traces which the plane field cuts on 2-dimensional surfaces. Suppose F is a 2-dimensional surface in the 3-manifold M. If F is transverse to ξ then the line field $T(F) \cap \xi$ on F integrates into a 1-dimensional foliation F_ξ which is called *characteristic*. Generically, F and ξ have isolated points of tangency, and thus the characteristic foliation F_ξ still can be defined as a foliation with isolated singularities.

In most of this paper we assume for simplicity that ξ is *co-orientable* or *transversely orientable, i.e.* that the normal bundle to the plane field is orientable. (The non-co-orientable case is discussed briefly in Section 2.9.1 below). A non-co-orientable tangent plane field admits a co-orientable double cover, so that many of the results proven in the co-orientable case can be automatically re-proven in the non-co-orientable one. Of course, for local considerations this is irrelevant.

We will also assume that the manifold M is compact, although in most cases the results remain true in the non-compact case.

Suppose we are given a triangulation of an oriented manifold M by small simplices such that each 1- and 2-dimensional simplex of the triangulation is transverse to ξ (compare [55]). Thus the characteristic foliation on each 2-simplex is non-singular. Each 3-simplex σ of the triangulation has exactly two vertices p and q such the ξ_p and ξ_q are supporting planes for the simplex. Let T be the edge connecting p and q. Let us pick any co-orientation of ξ. This co-orientation defines an orientation of T, and together with the orientation of M it defines an orientation of ξ, which gives an orientation of the characteristic foliation $(\partial\sigma)_\xi$. The holonomy along the leaves of the characteristic foliation $(\partial\sigma)_\xi$ defines a diffeomorphism $h_\sigma : T \to T$ (see Figure 1.1).

---

[1]This usage of "integrable" is the same as in multivariable calculus and differential equations. In suitable local coordinates, integral surfaces are graphs $(x, y, f(x, y))$ of functions f that satisfy ("integrate") a system of first-order partial differential equations, namely $f_x(x, y) = F(x, y, f(x, y))$ and $f_y(x, y) = G(x, y, f(x, y))$.

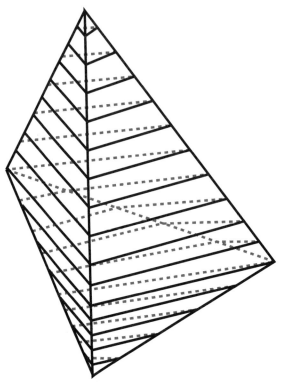

FIGURE 1.1. Characteristic foliation on the boundary of a 3-simplex: contact case.

If ξ is a foliation then all the holonomy maps h_σ equal the identity. For a general ξ this is, of course, not usually true, and it is important to consider two special cases:

a): $h_\sigma|_{\mathrm{int}\,T}$ is strictly decreasing (resp. increasing);
b): h_σ is non-strictly decreasing (resp. increasing).

REMARK 1.1.1. Notice that by changing the orientation of ξ we reverse simultaneously the orientations of T and $(\partial\sigma)_\xi$. Thus the property of h_σ to be decreasing or increasing is independent of the choice of the co-orientation of ξ.

Case a) corresponds to positive (resp. negative) contact structures, while case b) corresponds to positive (resp. negative) confoliations.

Let us now give precise definitions from the differentiable perspective. Suppose ξ can be defined (locally, if ξ is not co-orientable) by a Pfaffian equation $\{\alpha = 0\}$. The 1-form α is defined up to a multiplication by a non-vanishing function. The Frobenius theorem tells us that a necessary and sufficient condition for the integrability of ξ is the identity

$$\alpha \wedge d\alpha \equiv 0\,.$$

If $\alpha \wedge d\alpha$ does not vanish, then ξ is nowhere integrable and is called a *contact structure*. Notice that a manifold M which carries a contact structure is automatically oriented because the sign of the volume form $\alpha \wedge d\alpha$ is independent of the choice of

α. On the other hand, if the manifold M is à priori oriented (and we will always assume it for the rest of the paper) then one can distinguish between *positive and negative* contact structures depending on whether $\alpha \wedge d\alpha > 0$ or $\alpha \wedge d\alpha < 0$.

A plane field $\xi = \{\alpha = 0\}$ on an oriented manifold M is called positive (or negative) *confoliation* if $\alpha \wedge d\alpha \geq 0$, or respectively $\alpha \wedge d\alpha \leq 0$. Thus confoliations include contact structures and foliations on the two ends of the scale.

It is a simple exercise (or, alternatively follows from 1.3.4 below) to show that contact structures and foliations (and confoliations) can be constructed in an essentially unique way from the data of the holonomy maps maps h_σ subject to the appropriate inequalities as described above.

REMARK 1.1.2. (*Class of smoothness of a confoliation*) In this paper the class of smoothness of a confoliation is understood as the class of smoothness of the corresponding plane field (which is assumed to be at least C^1). On the other hand, in the theory of foliations the class of smoothness is usually understood as the class of smoothness of transition maps which form the corresponding cocycle. A foliation which is C^k-smooth in the second sense is C^k-smooth in the first one, but possibly for a different (but equivalent) smooth structure on the manifold. Conversely, C^k-smoothness of a foliation in the first sense implies its C^k-smoothness in the second one.

1.1.2. Local models. Both contact structures and foliations are locally homogeneous and admit a local normal form: foliations can be locally defined by the equation $dz = 0$, and contact structures by the equation $dz - ydx = 0$ (Darboux' theorem). In the case of negative contact structures it requires an orientation reversing change of coordinate to get this normal form. Confoliations, of course, need not to be locally homogeneous. Contact structures exhibit additional homogeneity properties. First, the group of contactomorphisms acts transitively on any connected contact manifold. Second, on a closed manifold there are no deformations of contact structures (see [**29**]), *i.e.* two contact structures which are homotopic as contact structures are isotopic. In contrast, there are continuous families of pairwise non-equivalent foliations, even infinite-dimensional families.

REMARK 1.1.3. The characterization of confoliations by holonomy maps $h|_\sigma$ gives a definition of confoliations that is more combinatorial than the definition by differential forms, although this definition is still not strictly combinatorial, since the holonomy maps involve infinite choices. Since foliations and confoliations have a high degrees of variability, often infinite-dimensional, a strictly combinatorial specification is not to be expected in those cases.

QUESTION 1.1.4. *Is there a more purely combinatorial theory for contact structures transverse to a triangulation? Is there a theory analogous to the normal surfaces in 3-manifolds, in which an isomorphism class of contact structures is specified by finite data subject to finitely verifiable conditions?*

1.1.3. Propeller. The following interpretation of the contact and confoliation conditions is quite useful. Suppose that a 3-manifold M splits as a product $F \times \mathbb{R}$ of a 2-surface F and the line $\mathbb{R}$, and the plane field ξ is tangent to the lines $q \times \mathbb{R}, q \in M$. The family of the characteristic foliations $\chi^t = (F \times t)_\xi$ on the surfaces $F_t, t \in \mathbb{R}$,

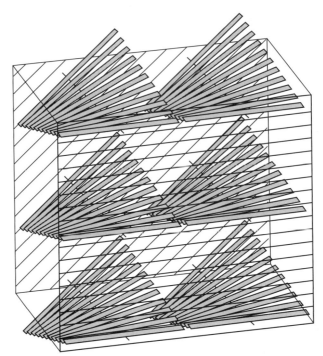

FIGURE 1.2. Characteristic foliations on two of the surfaces F_t are shown as the front and back faces of the figure. Going from front to back, the leaves of the characteristic foliations turn like propellers.

completely determines ξ. Thus ξ can be thought of as a line field on a surface F which varies in time.

Each foliation χ^t, $t \in \mathbb{R}$, can be defined by a non-vanishing 1-form α_t on F. The condition that ξ is a positive contact structure then reads as

$$(1.1) \qquad \alpha_t \wedge \frac{d\alpha_t}{dt} > 0,$$

which can be interpreted intuitively to mean that for any point $p \in F$ the line χ^t_p through this point keeps rotating in the positive direction (see Figure 1.2)[2]. Similarly, for a confoliation, the lines through every point should rotate in at most one direction, like a ratchet. They can stop somewhere and sometimes, but they can never turn backward.

Notice that locally any plane field is diffeomorphic to a field which is tangent to a parallel line field, and therefore it has the form considered above. Moreover,

---

[2]Warning: Note that this discussion does *not* involve a Riemannian metric, and in fact, this meaning of "rotation" differs from the most straightforward definition on a Riemannian manifold. Rotation in the current sense is not measured as deviation from parallel transport of the plane field. Instead, the intuitive image is that any leaf of a characteristic foliation describes a string of propellers, and as the slice F changes, the individual propeller blades rotate clockwise relative to the motion of the string.

The measure of non-integrability in terms of Riemannian geometry is that, given a 2-plane field, at each point you consider its average rate of twisting in directions tangent to the plane field. To be a positive contact structure, this average should be positive at every point.

one can choose a local coordinate system (x, y, z) so that z-curves are transversal to ξ while y-curves are tangent to ξ. Then the plane field ξ can be defined in these coordinates by a 1-form

(1.2) $$\alpha = dz - a(x, y, z)dx \,,$$

with

$$\alpha \wedge d\alpha = \frac{\partial a}{\partial y}(x, y, z)dx \wedge dy \wedge dz \,.$$

Thus we have

PROPOSITION 1.1.5. *Suppose a plane field ξ is defined by the differential form (1.2).*

Then

- ξ *is a foliation if and only if the function a is independent of y;*
- ξ *is a positive (resp. negative) contact structure if and only if*

$$\frac{\partial a}{\partial y} > 0 \quad (resp. \ \frac{\partial a}{\partial y} < 0);$$

- ξ *is a positive (or negative) confoliation if the function a is non-strictly increasing (or decreasing).*

Notice that coordinates (x, z, y) defines the negative orientation of $\mathbb{R}^3$. Thus 1.1.5 shows that planes which form a positive confoliation rotate *counter-clockwise* with respect to the standard orientation, when moving along the y-axis.

In a more global setting we have

PROPOSITION 1.1.6. *Suppose (M, ξ) admits a complete flow $X^t : M \to M, t \in \mathbb{R}$, generated by a non-vanishing vector field X tangent to ξ. Let η be the plane field orthogonal to X for a Riemannian metric on M. We orient η in such a way that X and η define the given orientation of M. Set $\lambda = \xi \cap \eta$, and denote by $\lambda_p^t, t \in \mathbb{R}$, the line $d_{X^t(p)}X^{-t}(\xi_{X^t(p)}) \cap \eta$, where $d_{X^t(p)}X^{-t} : T_{X^t(p)}(M) \to T_p(M)$ is the differential of the map X^{-t} at the point $X^t(p)$. Let $\theta_p(t)$ be the angle between the lines λ_p^0 and λ_p^t. Then ξ is a positive (resp. negative) contact structure if and only if*

$$\frac{d\theta_p(t)}{dt} < 0 \quad (resp. \ > 0)$$

for all $p \in M, t \in \mathbb{R}$. Similarly the confoliation condition amounts to the non-strict inequality.

1.1.4. Giroux' criterion. Let us recall that the *space of contact elements* $PT^*(V)$ of a manifold V is its projectivized cotangent bundle. For a surface F a point of $PT^*(F)$ is a tangent line to F, and $PT^*(F)$ is a circle bundle over F. If F is oriented then the fibration $\pi : PT^*(F) \to F$ is an oriented circle bundle over F with the Euler number $e = 2\chi(F)$.

The manifold $PT^*(F)$ carries a canonical contact structure η, tangent to the fibers. The contact plane η_z at a point $z = (x, l)$, $x \in F, l \in PT_x^*(F)$, projects by $d\pi : T(PT^*(F)) \to T(F)$ to the line $l \subset T(F)$. Ice skating can help form a good mental image for this contact structure: a skate can go forward and backward, and it can turn, but the blade does not (generally) slide sideways. The blade can be thought of as describing a contact element; its 2-dimensional space of possible motions from any position define the contact structure. Any fiber describes a motion of the skater spinning in place. (Strictly speaking, you should forget the distinction between front and back, for the ice skate to represent a contact element rather than a unit vector)

Notice that this contact structure is non-co-orientable, and that the orientation of $PT^*(F)$ defined by this structure is opposite to the one which is determined by the orientation of the surface F and the corresponding orientation of the fiber (compare the remark after Proposition 1.1.5).

The following observation of E. Giroux gives a necessary and sufficient condition for a circle fibration over a surface to carry a contact structure tangent to the fibers.[3]

PROPOSITION 1.1.7. (E. Giroux, [25]; comp. Arnold–Givental, [2], Section 2.1) *Let $M \to F$ be an oriented S^1- fibration over an orientable surface F. Then the equality*

$$(1.3) \qquad\qquad ke = 2\chi(F),$$

where e is the Euler number of the fibration $M \to F$, and $k \neq 0$ is an integer, is necessary and sufficient for the existence of a contact structure on M, tangent to the fibers of the fibration.

PROOF. Let ξ be a contact structure on M tangent to the fibers. The differential dp of the projection $p : M \to F$ maps contact planes onto tangent lines to F, and thus defines a map $q : M \to PT^*(F)$. In view of 1.1.6 the map q is the fiberwise covering map between the fibrations $M \to F$ and $PT^*(F)$, and the equality 1.3 follows. Conversely, the equality 1.3 implies existence of the fiberwise covering map $q : M \to PT^*(F)$, and the required contact structure ξ on M is the pull-back of the canonical contact structure η on $PT^*(F)$.   □

REMARK 1.1.8.   **1):** Notice that a *foliation* tangent to an S^1-fibration exists only in the case when F is the 2-torus T^2. On the other hand, the same equality (1.3) is necessary and sufficient for existence of a *confoliation*, *different from a foliation* tangent to the fibers. In particular, any non-trivial fibration over T^2 has a foliation tangent to the fibers, but does not admit a confoliation other than a foliation that is tangent to the fibers.[4]

2): It is interesting to notice that when $F \neq T^2$ then the condition (1.3) coincides with the pure homotopical condition for the existence of *any* plane field tangent to the fibers of the S^1-fibration. Moreover, the homotopy class of such a plane field on a fixed fibration is unique.

3): If the Euler number of a fibration satisfies the condition (1.3) with positive (resp. negative) k then the fibration admits a positive (resp. negative) contact structure tangent to the fibers.

Another interpretation of integrability comes from the complex world.

1.1.5. Pseudo-convexity.
Let M be an oriented 3-manifold endowed with a (co-orientable) tangent plane field ξ. Let us choose a complex structure J on the 2-dimensional bundle ξ. The pair (ξ, J) is called a *CR-structure* on M (CR stands for Cauchy-Riemann). The CR-structure is called

---

[3]A similar result for Seifert fibrations is proven by K. Mishachev in [45].

[4]V.I. Arnold ([1]) asked when there exists a contact structure which is tangent to a given vector field on a 3-manifold. Proposition 1.1.7 and Proposition 2.2.4 below provide partial information in this direction.

- *strictly pseudo-convex* if ξ is a positive contact structure,
- *pseudo-convex* if ξ is a positive confoliation, and
- *Levi-flat* if ξ is a foliation.

We will also use the terminology *J-convex, weakly J-convex*, and *J-flat* instead of strict pseudo-convexity, pseudo-convexity and Levi-flatness, respectively.

Given a 2-dimensional complex manifold W, we will call an embedding f of a CR-manifold (M, ξ, J) into W a *CR-embedding* if $df|_\xi$ is a complex linear map $\xi \to T(W)$. Thus a CR-manifold (M, ξ, J) admits a CR-embedding into a 2-dimensional complex manifold W if it can be realized as a hypersurface in W such that the field of complex tangent lines to the hypersurface coincides with the (ξ, J). Clearly, pseudo-convexity of a hypersurface is a property invariant under bi-holomorphic transformations.

PROPOSITION 1.1.9. (See, for instance, [**36**]) *Any real analytic CR-structure admits a CR-embedding into a 2-dimensional complex manifold.*

COROLLARY 1.1.10. *Given any strictly pseudo-convex, C^k-smooth $(k \geq 1)$ CR-structure (M, ξ, J), one can C^k-approximate J by J' such that the CR-structure (M, ξ, J') admits a CR-embedding into a complex manifold.*

PROOF. First approximate (M, ξ, J) by a real analytic CR-structure $(M, \tilde\xi, \tilde J)$, which will be automatically strictly pseudo-convex. According to 1.1.9 it is embeddable into a complex manifold. Then observe that there exists (see [**29**]) a diffeomorphism h that is C^k-close to the identity sending ξ to $\tilde\xi$. The CR-structure $(M, \xi, J' = h^* \tilde J)$ has the required properties. □

We do not know if Corollary 1.1.10 still holds for non-strictly pseudo-convex CR-structures.

REMARK 1.1.11. It is a subtle question whether a strictly pseudo-convex CR-structure can be realized as the boundary of a *compact* complex manifold (with boundary). As it is explained in Section 3.2 below, there exist contact manifolds (M, ξ) which do not admit any J such that the CR-manifold (M, ξ, J) can be realized as a boundary of a compact complex manifold.

The next proposition clarifies the geometric meaning of pseudo-convexity.

PROPOSITION 1.1.12. *Suppose the oriented CR-manifold (M, ξ, J) is realized as a hypersurface of a complex manifold W. Let us choose a Kählerian metric in a neighborhood U of a point $q \in M$. Let ν be the unit normal vector field to $M \cap U$ in W that is pointed to the semi-neighborhood of $M \cap U$ which induces on M the given orientation. Then $M \cap U$ is pseudo-convex (resp. strictly pseudo-convex, or Levi-flat) if and only if the mean normal curvature of $M \cap U$ in each of its tangent complex lines (i.e. in the direction of ξ) is non-negative (resp. positive, or identically zero). In other words, if Π is the second quadratic form of the hypersurface $M \cap U$ with respect to the chosen metric and the normal field, then the CR manifold is pseudo-convex (resp. strictly pseudo-convex, or Levi-flat) if and only if*

$$\Pi(v) + \Pi(Jv) \geq 0 \quad (resp. \; > 0, \; or \equiv 0)$$

for any non-zero vector $v \in \xi$.

The strict pseudoconvexity can also be characterized by the following property.

PROPOSITION 1.1.13. (See, for instance, [31]) *A hypersurface in a complex manifold W is strictly pseudoconvex if and only if in a neighborhood of each of its points it can be made strictly convex by a bi-holomorphic transformation.*

Of course, a bi-holomorphic image of a geometrically convex hypersurface is pseudo-convex. However, the converse is wrong without the assumption of strict pseudo-convexity.

Notice that one can also speak about pseudo-convex hypersurfaces in almost complex manifolds.

Few examples. Here are few examples of pseudo-convex hypersurfaces in complex 2-manifolds.

1. Recall that a surface is called *totally real* if none of its tangent planes is a complex line. Let F be a curve, or a totally real embedded surface in a complex manifold (W, J). Then the oriented boundary of a sufficiently small tubular neighborhood N of F is J-convex. This is true for any choice of Riemannian metric on W. The geometric reason is that, since there are no complex lines tangent to F, any complex line tangent to the boundary of a tubular neighborhood of F must have a (real) component that points circumferentially, in a direction that is strongly curved for a small tubular neighborhood.

2. Suppose that a totally real surface $F \subset W$ is immersed with transversal intersections, so that for each intersection point the union $\tau_1 \cup \tau_2$ of the tangent planes to the intersecting branches is invariant under the action of J. Then F has an arbitrarily small regular neighborhood with J-convex boundary. However, in this case the shape of N near the intersection points should be chosen in a special way (see [15]).

3. Let M be a J-convex boundary of a domain $\Omega \subset W$, and $F \subset W$ a totally real surface. Suppose that

- F intersects M transversely along its boundary ∂F;
- ∂F is *Legendrian*, *i.e.* for any point $p \in \partial F$ we have $JT_p(F) \subset T_p(M)$;
- F is attached to M from the outside, *i.e.* for any point $p \in \partial F$ the interior normal vector to ∂F in F is an outside transversal to the boundary M of the domain Ω.

Then $\Omega \cup F$ has arbitrarily small neighborhood with J-convex boundary (see [15]).

1.1.6. Confoliations on manifolds of dimension > 3.
The following high-dimensional generalization of the notion of confoliation seems reasonable in view of the complex-analytic analogy considered in the previous section. However, we do not have any non-trivial results concerning this object.

A tangent hyperplane field $\xi = \{\alpha = 0\}$ on a $(2k + 1)$-dimensional manifold M is called a *positive confoliation* if there exists a complex structure J on the bundle ξ such that

$$d\alpha(X, JX) \geq 0$$

for any vector $X \in \xi$. Thus the confoliation condition for ξ is by definition equivalent to the existence of a compatible weakly pseudo-convex CR-structure.

1.2. Dynamics of codimension one foliations

1.2.1. Basic notions of the theory of codimension one foliations. Although we are concerned in this paper only with 3-dimensional manifolds, all the results which we discuss in this section hold for codimension 1 foliations on manifolds of arbitrary dimension.

The Reeb stability theorem claims that a co-orientable foliation on a closed manifold which contains a closed simply connected leaf is a fibration over a circle. In particular, in the three-dimensional (orientable) case such foliation has to be diffeomorphic to the foliation of $S^2 \times S^1$ by spheres $S^2 \times p$, $p \in S^1$. We will assume without further notice that the foliations which we consider *have no simply-connected closed leaves*.

Let ξ be a co-oriented codimension 1 foliation on a closed manifold M. Given an oriented immersed curve Γ which is contained in a leaf S of the foliation, we define the *holonomy* along Γ as a germ at 0 of a diffeomorphism $\mathbb{R} \to \mathbb{R}$ as follows. Take an annulus $A = \Gamma \times [-1, 1]$ along $\Gamma = \Gamma \times 0$ which is transversal to ξ. Then ξ defines on A the characteristic foliation A_ξ which has Γ as its closed leaf. The holonomy is the germ of the corresponding Poincaré return map which is well defined and depends only on the homotopy class of the loop Γ in S.

A *foliation without holonomy* is a foliation for which the holonomy along any curve is trivial. We say that the holonomy along Γ is *non-trivial* if it is different from the identity. If the holonomy satisfies $|f(x)| < |x|$ for x sufficiently near 0 but $\neq 0$, we say the holonomy is *attracting*. This is equivalent to the condition that under iteration of f, all nearby points tend toward 0. The opposite inequality $(|f(x)| > |x|)$ is called *repelling*.

If there are intervals close to 0 in which $f(x) < x$, where f is the holonomy, we say f is *sometimes attracting*. A sometimes attracting germ which also satisfies the inequality $|f(x)| \leq |x|$ for all $x \neq 0$ sufficiently close to 0 is *weakly attracting*. If the foliation is co-oriented, these definitions can be restricted to the positive values of x, and we say f is *sometimes attracting on the positive side*, and so on. We can also reverse the inequalities for f in these definitions and replace "attracting" by "repelling".

We say that Γ has a non-trivial *linear holonomy* if the differential of the holonomy map at 0 is different from 1.

REMARK 1.2.1. If a holonomy curve Γ has non-trivial holonomy then arbitrarily close to Γ one can find another curve (perhaps on a different leaf) which has attracting holonomy on one of its sides.

1.2.2. Structure of foliations without holonomy. Let Y be a 1-manifold and N be a manifold. A *foliated Y-bundle* over a manifold N is a pair (W, ξ) where W is a fibration over N with a fiber Y and ξ is a codimension 1 foliation on W which is transversal to the fibers of the fibration.

If Y is compact then foliated Y-bundles over N are classified by homomorphisms ρ from $\pi_1(N)$ to the group of diffeomorphisms of Y. For the foliated bundle to be without holonomy is equivalent to the statement that whenever $\rho(\alpha)$ has a fixed point, it is the identity. Since there are no fixed points, this means that the diffeomorphisms in question are circularly ordered, and a careful analysis shows that they must commute with each other (compare [**48**]). In general,

THEOREM 1.2.2. *(See* [33] *and* [47]*) Let* (M, ξ) *be a foliation without holonomy. Then there exists a foliated trivial bundle* $(T^n \times S^1, \eta)$ *without holonomy and a map* $h : M \to T^n \times S^1$ *transversal to* η *which induces* ξ *from* η.

A subset of a foliation is *saturated* if whenever one point on a leaf is in the set, then the entire leaf is contained in the set. A *minimal set* in a foliated manifold (M, ξ) is a non-empty closed saturated subset which does not contain any smaller non-empty closed saturated subsets. In other words, a set is minimal if it is the closure of any leaf. In any compact manifold, each closed saturated subset contains at least one minimal set. A minimal set is called *exceptional* if its intersection with any transversal curve is a nowhere dense Cantor set.

The most obvious examples of foliations without holonomy are foliations of $\mathbb{R}^{n+1}$ by planes, modulo $\mathbb{Z}^{n+1}$ acting as a lattice. Denjoy [5] invented a construction of "blowing up" leaves that can be used to modify these to give C^1 foliations without holonomy having exceptional minimal sets. The idea is to choose a single leaf, and replace it by an intervals worth of leaves, in a careful way so that the resulting foliation is still C^1. (Actually Denjoy did this on T^2, but the general case follows readily). However, Denjoy also showed that this construction cannot be made C^2.

COROLLARY 1.2.3. *If the foliated manifold* (M, ξ) *is as in Theorem 1.2.2 and* C^2*-smooth then it can be* C^0*-approximated by a fibration over a circle.*

PROOF. According to [5] there exists a homeomorphism $g : S^1 \to S^1$ which conjugates ρ to a group of rotations of the circle, *i.e.*, the foliation is topologically conjugate to a linear foliation η_0 on the torus $T^{n+1} = T^n \times S^1$. We can C^0-approximate g by a diffeomorphism $\tilde{g}$, and approximate η_0 by a linear foliation $\tilde{\eta}_0$ with a rational slope. Then $\tilde{\eta}_0$ is a fibration $p : T^{n+1} \to S^1$. If the chosen approximations are sufficiently good then the map h constructed in 1.2.2 is still transversal to the foliation $\tilde{\eta} = (\mathrm{id} \times \tilde{g})^*(\tilde{\eta}_0)$, and therefore the foliation $\tilde{\xi} = h^*(\tilde{\eta})$ which is C^0-close to ξ is the foliation by the fibers of the fibration $p \circ (\mathrm{id} \times \tilde{g}) \circ h : M \to S^1$. $\qquad \square$

The *center* $Z(\xi)$ of ξ is the union of all its minimal sets. The union of all closed leaves of ξ will be denoted by $C(\xi)$.

THEOREM 1.2.4. *(see* [33]*) Let* A *be a minimal set. Then* A *is either the whole manifold* M, *or a closed leaf, or an exceptional minimal set. The sets* $C(\xi)$ *and* $Z(\xi)$ *are closed. Each exceptional minimal set is isolated in* $Z(\xi)$, *i.e. has a neighborhood which contains no other minimal sets. In particular, there are only finitely many exceptional minimal sets.*

The previous theorem holds without any assumptions about the smoothness of the foliation. The next theorem, due to R. Sacksteder, requires that the foliation is at least C^2-smooth.

THEOREM 1.2.5. *(see* [51]*) Any exceptional minimal set of a* C^2*-foliation contains a curve with non-trivial linear holonomy.*

Theorem 1.2.5, in fact, is a corollary of another Sacksteder's theorem concerning finitely generated pseudogroups of C^2-diffeomorphisms between open intervals of a line.

THEOREM 1.2.6. (see [51]) *Let $\mathcal{P}$ be a finitely generated pseudogroup of C^2-diffeomorphisms between open intervals of a line. Suppose it has an exceptional, i.e. minimal invariant Cantor set $C \subset \mathbb{R}$. Then there exists an element $g \in \mathcal{P}$ and a point $x \in C$ such that $g(x) = x$ and $g'(x) < 1$.*

E. Ghys ([26]) has observed that Theorem 1.2.6 implies existence of linear holonomy even in the situation when the foliation is *minimal, i.e.* all its leaves are everywhere dense.

THEOREM 1.2.7. (E. Ghys) *If a minimal C^2-foliation has non-trivial holonomy then it has non-trivial linear holonomy.*

REMARK 1.2.8. The remaining case, when the foliation is not minimal and when it does not have exceptional minimal sets, is that the only minimal sets are closed leaves. The Reeb foliation of S^3 is the classical example of a foliation with a single closed leaf which does not, and can not, have nontrivial linear holonomy around its closed leaf.

The result follows from Theorem 1.2.6 and the following lemma applied to the holonomy pseudogroup of the foliation.

LEMMA 1.2.9. (E. Ghys) *Suppose a pseudogroup $\mathcal{P}$ of homeomorphisms between intervals of $\mathbb{R}$ has dense orbits. Then if $\mathcal{P}$ has non-trivial holonomy (i.e. there exists an element $g \in \mathcal{P}$ and a point $x \in \mathbb{R}$ such that $g(x) = x$ and the germ of g at x is different from the identity) then it has a finitely generated subpseudogroup $\mathcal{P}_0$ which has an exceptional minimal set.*

PROOF. First observe that there exists an element $f \in \mathcal{P}$ which is defined on an interval $I = [a, b]$ and satisfies the conditions

$$f(a) = a, \quad \text{and} \quad f(x) < x \quad \text{for all} \quad x \in (a, b].$$

Since the orbits of $\mathcal{P}$ are dense, there exists another $g \in \mathcal{P}$ which is defined on $[a, a']$ and such that $a < g(a) < b$. Taking a smaller a', if necessary, we can assume that

$$a' < g(a) \quad \text{and} \quad g(a') < b.$$

Replacing f by a power of f we can further assume that $f(b) < a'$. Thus we have

$$a < f(b) < a' < g(a) < g(f(b)) < b.$$

Let $\mathcal{P}_0$ be the subpseudogroup of $\mathcal{P}$ generated by $f_1 = f$ and $f_2 = g \circ f$. Both diffeomorphisms f_1 and f_2 are defined on the interval $I = [a, b]$ and map it onto two disjoint subintervals $I_1 = f_1([I]) = [a, a'']$, where $a'' < f(b) < a'$, and $I_2 = f_2(I) = [g(a), g(f(b))]$. Define now the sequence of closed subsets

$$I = C_0 \supset C_1 \supset C_2 \supset \ldots$$

by the recursive formula

$$C_0 = I \quad \text{and} \quad C_k = f_1(C_{k-1}) \cup f_2(C_{k-1}) \quad \text{for} \quad k = 1, \ldots .$$

Then the Cantor set $C_\infty = \bigcap_i C_i$ is a minimal invariant set for the subpseudogroup $\mathcal{P}_0$.   □

We finish this section with the following property of the set $C(\xi)$, proved in [6]. We consider here only the 3-dimensional case, which is only needed for our purposes, and refer the reader to [6] for the general case.

PROPOSITION 1.2.10. *Any C^k-foliation of a closed 3-manifold, different from the foliation of $S^2 \times S^1$ by spheres $S^2 \times p$, $p \in S^1$, (see the remark at the beginning of Section 1.2). can be C^k-approximated by a foliation with only a finite number of closed leaves.*

PROOF. Take a closed leaf S. Then it is either isolated in $C(\xi)$, or at least on one of its sides there are other closed leaves, arbitrarily close to S. In the latter case the holonomy map along any curve in S is C^k-flat at the origin. Let us denote by $\mathcal{A}$ the set of all non-isolated closed leaves. If the leaf S is an isolated point of $\mathcal{A}$ then we can insert instead of S a thin foliated I-bundle U over S which is C^k-close to the trivial horizontal foliation, has no interior closed leaves and has C^k-flat holonomy along curves in the boundary leaves. To do this one just need to choose a diffeomorphism $\varphi : I \to I$ without interior fixed points, which is flat at the boundary points. The required foliated I-bundle is determined by the representation $\pi_1(S) \to \mathrm{Diff}\,I$, which sends one of the generators of $\pi_1(S)$ to φ and all others to the identity. The new foliation has two isolated closed leaves instead of one leaf $S \in \mathcal{A}$. Suppose now that S is not an isolated point of $\mathcal{A}$. Then there exits another closed leaf $S' \in \mathcal{A}$ which is arbitrarily close to S. The leaves S and S' bound together a foliated I-bundle U' over S which is C^k-close to the trivial horizontal foliation. If the holonomy along S' is C^k-flat then we can delete this foliation and insert instead a foliated I-bundle U, described above. Notice that the foliation constructed by this surgery is C^k-smooth because the holonomy along S and S' is C^k-flat on both sides of the closed leaves. If the holonomy along S' is not flat then one first should flatten it using the following (compare [58])

LEMMA 1.2.11. *Let f be a germ at 0 of a C^k-diffeomorphism $\mathbb{R}_+ \to \mathbb{R}_+$, where $k \geq 2$. Set $g(x) = \exp(-\exp(1/x^2))$. Then the germ $h = g \circ f \circ g^{-1}$ is C^∞-flat at the origin.*

Thus, if we take a homeomorphism $F : M \to M$ which is fixed outside a small tubular neighborhood of the leaf S' and whose germ along S' equal $\mathrm{id} \times f$, then the foliation $F_*(\xi)$ is C^∞-flat along S' and C^k-close to ξ.[5] Repeating this procedure, if necessary, on the other side of S we destroy all closed leaves in a neighborhood of S without changing the structure of the set $C(\xi)$ outside of this neighborhood. The compactness of $C(\xi)$ guarantees us that we can get a foliation with isolated closed leaves in a finite number of steps.   □

1.2.3. Laminations. Let A be a closed subset of a 3-manifold M. A tangent plane field ξ on M given at the points of A is called a *lamination* if for each point $p \in A$ there exists a surface F and an embedding $f : F \to A$ which is tangent to ξ, contains the point p, and is complete in the following sense: for each point $q \in M$ there exists a closed ball B centered at q such that each connected component of the pre-image $f^{-1}(B)$ is compact. We say that ξ is a C^k-lamination if it extends to a C^k-smooth plane field on a neighborhood of A (compare Remark 1.1.2 above).

---

[5]Warning: the above flattenning trick does not improve the class of smoothness of the foliation along the closed leaf S.

The union of all integral surfaces of ξ forms a partial foliation $\mathcal{F}$ of M. Namely each point $q \in M$ admits a neighborhood U_q and a submersion $\pi_q : U_q \to \mathbb{R}$ such that the restriction $\mathcal{F}|_A$ is a foliation defined by pre-images of points of a closed set $C \in \mathbb{R}$.

The following proposition shows how laminations enter into the study of plane fields on 3-manifolds.

For any plane field ξ on a closed manifold M we denote by $\mathrm{Fol}(\xi)$ its *fully foliated part*, *i.e.* the set of points $q \in M$ such that there exists an embedding of a complete surface F into M tangent to ξ which contains the point q.

PROPOSITION 1.2.12. *For a C^1-smooth plane field ξ the set $\mathrm{Fol}(\xi)$ is closed.*

Thus the restriction of ξ to $\mathrm{Fol}(\xi)$ is a lamination.

PROOF. Fix a Riemannian metric in M. It is sufficient to prove that for each point $p \in \overline{\mathrm{Fol}(\xi)}$ there exists an $\varepsilon > 0$ such that p is contained in an integral surface S_p such that $S_p \cap B_\varepsilon(p)$ is a closed submanifold of the ball $B_\varepsilon(p)$ of radius ε, centered at the point p. This is true, by the definition, for $p \in \mathrm{Fol}(\xi)$. Take a boundary point $p \in \partial \mathrm{Fol}(\xi)$, and let $p_k, k = 1, \ldots$, be a sequence of points of $\mathrm{Fol}(\xi)$ which converges to p. Notice that in view of the compactness of M, the normal curvatures of any integral surface of the plane field ξ are a priori bounded by a constant, which depends only on ξ. Let $D_k, k = 1, \ldots$, be the intersection of the leaf of ξ through the point p_k with the ball $B_\varepsilon(p)$. If ε is sufficiently small and k is large then D_k is a disc. The sequence of these discs converges to a C^1-smooth disc D through the point p which is automatically integral for ξ. Notice that a posteriori we can conclude that the disc is as smooth as the plane field ξ. □

The results from the previous section about the foliations with-holonomy hold with practically no changes for laminations. Notice that the notion of non-trivial linear holonomy has sense even for the laminations. We say that a lamination (A, ξ) has non-trivial linear, say attracting holonomy along an integral curve Γ if, given a transversal curve T through a point $p \in \Gamma$, the point p is an accumulation point of $A \cap T$ from both sides, and the germ h at p of the holonomy map $A \cap T \to A \cap T$ satisfies the inequality $h(x) < cx$ for all $x \in A \cap T$ for a positive constant $c < 1$.

The following proposition contains a summary of the results about the dynamics of laminations, which we need for our purposes.

PROPOSITION 1.2.13. *Let (A, ξ) be a C^k-lamination, $k > 1$. All minimal sets of the lamination ξ are either closed leaves or exceptional sets. The union $Z(\xi)$ of all minimal sets and the union $C(\xi)$ of all closed leaves are compact. All exceptional minimal sets are isolated and each of them contains a curve, along which ξ has non-trivial linear holonomy.*

As in the case of foliations, only the last statement about linear holonomy requires an assumption about the smoothness of ξ.

1.3. Plane fields transversal to 1-dimensional bundles

Let $\pi : M \to F$ be a fibration of a 3-manifold over a surface F with a one-dimensional fiber Y, and ξ a plane field transversal to the fibers of the fibration. We will be interested in three cases:

A): $M = F \times \mathbb{R}_+$ and ξ is the germ of a plane field tangent to the zero section $F \times 0$;

B): $M = F \times \mathbb{R}$ and ξ is a product-foliation outside a compact set.

C): $Y = S^1$

In the case A) we will study the local structure of the plane field near the zero-section and will denote by Ξ the space of germs along F of confoliations on M which have F as an integral leaf.

The plane field ξ can be viewed as a G-connection on this bundle, where G is the group $\mathrm{Diff}_+ S^1$ of orientation preserving diffeomorphisms of S^1 in the case C), G is the group $\mathrm{Diff}_0 \mathbb{R}$ of germs at the origin of diffeomorphisms $\mathbb{R}_+ \to \mathbb{R}_+$ in the case A), and G is the group $\mathrm{Diff}_{\mathrm{comp}} \mathbb{R}$ of compactly supported diffeomorphisms $\mathbb{R} \to \mathbb{R}$ in the case B).

The parallel transport along a curve $\Gamma \subset F$ with the end-points p and q is just a holonomy diffeomorphism $Y_p = \pi^{-1}(p) \to Y_q = \pi^{-1}(q)$ along the leaves of the characteristic foliation defined by ξ on the cylinder $Y_\Gamma = \pi^{-1}(\Gamma)$. The curvature Ω of this connection is a differential 2-form on F valued in the Lie algebra $\mathcal{V}$ of vector fields on Y (or germs of the vector fields at the origin, in the case A). As usual, it associates with a pair of vectors $T_1, T_2 \in T_p(F)$, $p \in F$, the holonomy along the boundary of an infinitesimal parallelogram generated by these vectors. Equivalently, one can define $\Omega(T_1, T_2)$ as follows. Take any two vector fields $\tilde{T}_1$ and $\tilde{T}_2$ on F in a neighborhood of the point p which extend T_1 and T_2, and lift them to the vector fields $\hat{T}_1$ and $\hat{T}_2$ in a neighborhood of the fiber Y_p, tangent to ξ, so that $\hat{T}_1, \hat{T}_2$ project to T_1 and T_2 by $d\pi$. Then $\Omega(T_1, T_2)$ is the Y-component of the Lie bracket[6] $[\hat{T}_1, \hat{T}_2]$. Notice that if ξ is defined by the Pfaffian equation $\{\alpha = 0\}$ then $d\alpha(\hat{T}_1, \hat{T}_2) = -\alpha([\hat{T}_1, \hat{T}_2])$. Thus the vector field $\tau = \Omega(T_1, T_2)$ on Y_p is defined by the equation

$$\alpha(\tau(y)) = -d\alpha(\hat{T}_1, \hat{T}_2), \ y \in Y_q.$$

The last formula implies the following characterization of confoliations transversal to the fibration $\pi : M \to F$ in terms of their curvature.

PROPOSITION 1.3.1. *Suppose that the projection $d\pi : \xi \to T(F)$ preserves the orientation, and Y is oriented according to the co-orientation of ξ. Let linearly independent vectors $T_1, T_2 \in T_p(F)$, $p \in F$, define the given orientation of F. Then ξ is a positive contact structure (resp. confoliation) if and only if the vector field $\Omega(T_1, T_2)$ does not vanish and defines the negative orientation of Y (resp. defines a negative orientation of Y at the points where it does not vanish) for all $p \in F$ and $T_1, T_2 \in T_p(F)$. The plane field ξ is a foliation if and only if $\Omega \equiv 0$.*

REMARK 1.3.2. This discrepancy in the terminology: *positive confoliations have negative curvature* is unfortunate but unavoidable.

If $Y = S^1$, and the fibration over an oriented curve Γ is trivialized, then the holonomy map $Y_p \to Y_p$, $p \in \Gamma$, can be lifted to a covering diffeomorphism $\mathbb{R} = \tilde{Y}_p \to \tilde{Y}_p$ which depends only on the homotopical class of the trivialization. When the trivialization is fixed we will always view the holonomy as a covering 1-periodic diffeomorphism $\mathbb{R} \to \mathbb{R}$. In particular, the trivialization, and therefore the lifting, is canonically defined if the curve Γ is boundary of an embedded disc $D \subset F$.

---

[6] We use the convention $[X, Y] = XY - YX$.

Fix a point $p \in \partial D$ and define the *(non-commutative) integral* $\int_{D,p} \Omega$ to be equal to the holonomy diffeomorphism $Y_p \to Y_p$ along the oriented boundary ∂D. In the case $Y = S^1$, as was already stated above, we define the integral $\int_{D,p} \Omega$ to be equal to the covering diffeomorphism $\mathbb{R} = \widetilde{Y}_p \to \widetilde{Y}_p$.

The definition is motivated by the following obvious properties of this integral.

LEMMA 1.3.3.   **a):** *Let P_ε be the image under the exponential map* $\exp : T_p(F) \to F$ *of the parallelogram in $T_p(F)$ generated by the vectors $\varepsilon T_1, \varepsilon T_2 \in T_p(F)$ (we fix a Riemannian metric on F). Then*

$$\lim_{\varepsilon \to 0} \frac{\int_{P_\varepsilon,p} \Omega - \mathrm{id}}{\varepsilon^2} = \Omega(T_1, T_2).$$

b): *(Additivity) Suppose a disc D is divided into two sub-discs D_1 and D_2 by an arc Γ which connects two points $p, q \in \partial D$, so that the positively oriented arc connecting p and q in ∂D is contained in ∂D_2. Then*

$$\int_{D,p} \Omega = \int_{D_1,p} \Omega \circ \int_{D_2,p} \Omega.$$

If the fibration $\pi : M \to F$ is trivial and the trivialization $M = F \times Y$ is fixed then we can view the plane field ξ as a connection form, still denoted by ξ, taking values in the Lie algebra $\mathcal{V}$ of vector fields on Y. Thus given a vector $\tau \in T_p(F)$ we have $\xi_p(\tau) \in \mathcal{V}$. Then $\Omega = d\xi$, and thus we have the "non-commutative Stokes theorem"

$$\int_{D,p} \Omega = \int_{\partial D,p} \xi.$$

Here the second integral is, by definition, the diffeomorphism $Y_p \to Y_p$ which is equal to the time 1 map of the flow $\varphi^t : Y \to Y, t \in I$, defined by the differential equation

$$\frac{d\varphi^t}{dt}(x) = \xi_{\gamma(t)}\left(\frac{d\gamma(t)}{dt}\right)(\varphi^t(x)),$$

where $\gamma : I \to \partial D$ is a parametrization of ∂D which defines the orientation of ∂D as the boundary of D.

LEMMA 1.3.4. *Let M, F, Y, π be as above and ξ a positive confoliation which is transversal to the fibers of the fibration $\pi : M \to F$. Then for any disc $D \subset F$ we have $\int_{D,p} \Omega \leq \mathrm{id}$, where the equality $\int_{D,p} \Omega = \mathrm{id}$ holds if and only if $\xi|_D$ is a foliation. If ξ is a contact structure then the strict inequality $\int_{D,p} \Omega < \mathrm{id}$ holds.*

PROOF. The inequality $\int_{D,p} \Omega \leq \mathrm{id}$ is straightforward from 1.3.3 and 1.3.1. If $\int_{D,p} \Omega = \mathrm{id}$ then $\Omega \equiv 0$ in D which means that $\xi|_D$ is a foliation. $\square$

Notice here that Lemma 1.3.4 explains the description of the contact and confoliation conditions given in Section 1.1 in terms of the holonomy maps h_σ.

Let us show now that the converse to 1.3.4 is also partially true.

PROPOSITION 1.3.5. *Let Y and the group G of its diffeomorphisms be as above. Suppose a diffeomorphism $\varphi : Y \to Y$ from G (or the covering 1-periodic diffeomorphism $\mathbb{R} \to \mathbb{R}$ from the universal covering group $\widetilde{G} = \widetilde{\mathrm{Diff}}_+ S^1$ in the case C)) satisfies the inequality $\varphi \leq \mathrm{id}$. Then there exists a positive confoliation ξ on $\mathrm{D} \times Y$ transversal to the fibers of the fibration $\pi : D \times Y \to D$ and such that the holonomy map along its positively oriented boundary equals φ. If $\Phi \subset Y$ is the set of fixed points of φ then the confoliation ξ can be constructed to have discs $D \times y, y \in \Phi$ as its integral leaves, and to be contact on $D \times (Y \setminus \Phi)$.*

PROOF. The diffeomorphism φ can be included into the flow φ_t generated by a vector field v_t on Y, 2π-periodic in t, so that $\varphi = \varphi_{2\pi}$ and φ_t satisfies the differential equation

$$\frac{d\varphi_t}{dt}(x) = v_t(\varphi_t(x)), \quad \varphi_0(x) = x.$$

The field v_t can be chosen in such a way that for all $t \in [0, 2\pi]$ we have $v_t(x) = 0$ if $x \in \Phi$ and $v_t(x) < 0$ for $x \in Y \setminus \Phi$. The required confoliation ξ can be defined by the 1-form $dz + \rho^2 v_\theta(z)d\theta$, where $z \in Y$ and $\rho \in [0,1], \theta \in [0, 2\pi]$ are polar coordinates in the disc D. $\qquad\square$

Let us look more closely at case B). We denote by G_k, $k = 0, \ldots, \infty$, the subgroup of germs at 0 of diffeomorphisms $\mathbb{R}_+ \to \mathbb{R}_+$ which are C^k-tangent to the identity at the origin, so that we have $G = G_0 \supset G_1 \supset \cdots \supset G_\infty$. We consider also the Lie algebra $\mathcal{V}_k$ of G_k which consists of germs at 0 of vector fields on $\mathbb{R}_+$ which have zero of order $> k$ at the origin. We set $H_k = G_k \setminus G_{k+1}$ and $\mathcal{W}_k = \mathcal{V}_k \setminus \mathcal{V}_{k+1}$. Notice that each of the spaces H_k and $\mathcal{W}_k$ consists of two components: $H_k = H_k^+ \cup H_k^-$ and $\mathcal{W}_k = \mathcal{W}_k^+ \cap \mathcal{W}_k^-$. The component H_k^- consists of attracting germs, the corresponding component $\mathcal{W}_k^-$ consists of germs of negative vector fields. Then similarly to 1.3.4 and 1.3.5 we can show:

PROPOSITION 1.3.6. *If the curvature Ω of a germ $\xi \in \Xi$ of a confoliation on $D \times \mathbb{R}_+$ takes values in $\mathcal{W}_k^+$ (resp. in $\mathcal{W}_k^-$) somewhere in D then the holonomy map $\int_{D,p} \Omega$, $p \in \partial D$, belongs to H_k^+ (resp. in $\mathcal{W}_k^-$). Conversely, for any germ φ from H_k^+ (resp. in $\mathcal{W}_k^-$) there exists a germ $\xi \in \Xi$ of a confoliation on $D \times \mathbb{R}^+$ whose curvature form Ω takes values in $\mathcal{W}_k^+$ (resp. in $\mathcal{W}_k^-$) everywhere in D, and*

$$\int_{D,p} \Omega = \varphi.$$

Let us denote by ζ the product-foliation (fibration) of the manifold $S^2 \times S^1$ by the spheres $S^2 \times z, z \in S^1$.

PROPOSITION 1.3.7.   **(a):** *Let ξ be a confoliation on $M = S^2 \times S^1$ which is transverse to the circles $u \times S^1, u \in S^2$. Then ξ is a foliation diffeomorphic to the fibration ζ.*

(b): *Let ξ be a confoliation on $M = S^2 \times \mathbb{R}$ which is transverse to the lines $u \times \mathbb{R}, u \in S^2$ and is tangent to the sphere $S = S^2 \times 0$. Then there exists a neighborhood U of the the sphere S such that $\xi|_U$ is a foliation diffeomorphic to the fibration of $S^2 \times I$ by the spheres $S^2 \times v, v \in I$.*

PROOF. Present S^2 as the union of two hemispheres $S^2 = D_+ \cup D_-$. Let φ be the covering holonomy $\mathbb{R} \to \mathbb{R}$ (or, in the case (b), the germ at 0 of the holonomy diffeomorphism $\mathbb{R} \to \mathbb{R}$) along the equator $E = \partial D_- = \partial D_+$ oriented as the boundary of D_+. Applying Lemma 1.3.4 to the disc D_+ we get the inequality $\varphi \leq \mathrm{id}$. On the other hand, the application of Lemma 1.3.4 to the disc D_- gives the opposite inequality. Thus $\varphi = \mathrm{id}$, and the same lemma implies that in the case a) ξ is a foliation diffeomorphic to ζ. In the case b) the above argument provides a diffeomorphism of the restriction of the foliation ξ to a neighborhood of S onto the restriction of the fibration ζ to a neighborhood of one of its fibers.   □

COROLLARY 1.3.8. *Any confoliation on $S^2 \times S^1$ which is C^0-close to the foliation ζ is a foliation, diffeomorphic to ζ.*

As was already mentioned above in Section 1.2.1, *Reeb stability theorem* for foliations claims that a co-orientable foliation which has a closed leaf diffeomorphic to S^2 has to be diffeomorphic to the foliation ζ on $S^2 \times S^1$. Now we can prove a similar result for confoliations.

PROPOSITION 1.3.9. *("Reeb stability theorem" for confoliations).   Suppose that a confoliation ξ on a closed oriented manifold M has an integral embedded 2-sphere S. Then ξ is a foliation and (M, ξ) is diffeomorphic to $(S^2 \times S^1, \zeta)$.*

PROOF. According to Proposition 1.3.7b) the restriction of ξ to a neighborhood of the leaf S is a foliation by spheres. Moreover, it follows that the subset F of M, where ξ is a foliation by 2-spheres, is open. On the other hand, according to 1.2.12 and 1.2.4 the set F is closed. Hence $M = F$ and ξ is a fibration by 2-spheres, which implies that (M, ξ) is diffeomorphic to $(S^2 \times S^1, \zeta)$.   □

COROLLARY 1.3.10. *Let ξ be a confoliation on the 3-ball B which is standard (i. e. coincides with the foliation of the round ball by planes) near the boundary. Then ξ is a foliation, and is diffeomorphic to the standard foliation of the round ball.*

In other words, one cannot convert a foliation into a confoliation by a local modification near a point.

PROOF. Consider an arbitrary confoliation (B, ξ) as in the statement of the corollary. Take the foliation ζ on $S^2 \times S^1$, and let B be a small ball which does not intersect one of the spheres $S^2 \times z_0$, $z_0 \in S^1$, such that the foliation induced on B is standard. Implant xi as a replacement for the foliation in B. The new confoliation is still tangent to the sphere $S^2 \times z_0$, and thus, according to 1.3.9, it is a foliation diffeomorphic to ζ. It follows then that the confoliation (B, ξ) is a foliation diffeomorphic to the standard one.   □

Let F be an oriented surface of genus $k > 0$. Consider the standard cell-decomposition of F with one 2-cell D, and the bouquet of circles

$$B = a_1 \vee b_1 \vee \cdots \vee a_k \vee b_k \subset F, \quad k = \mathrm{genus}(F),$$

be a bouquet of circles with the vertex $p \in F$ as its 1-dimensional skeleton. Thus we view D as $4n$-gon whose sides are labeled with

$$a_1, b_1, a_1^{-1}, b_1^{-1}, a_2, b_2, \ldots, a_k, b_k, a_k^{-1}, b_k^{-1},$$

and we assume that the circles a_i, b_i, $i = 1, \ldots, k$, are oriented by the attaching map $\partial D \to B$. We will identify the point $p \in B$ with the first vertex of the side a_1 of the $4n$-gon D.

PROPOSITION 1.3.11. *Let $\pi : M \to F$ be a fibration with the fibers $Y = S^1$. Set $e = e(\pi)[F]$, where $e(\pi)$ is the Euler class of the fibration $\pi : M \to F$. Suppose we are given any 1-periodic diffeomorphisms $f_1, g_1, \ldots, f_k, g_k \in \widetilde{G} = \widetilde{\mathrm{Diff}}_+(S^1)$. such that the product of their commutators*

$$h = [f_1, g_1] \circ \cdots \circ [f_k, g_k],$$

where $[f_i, g_i] = f_i \circ g_i \circ f_i^{-1} g_i^{-1}$, satisfies the inequality

(1.4) $$h(x) < x - e, \quad \text{for all} \quad x \in \mathbb{R}.$$

Then there exists a contact structure ξ transversal to the fibers of the fibration $\pi : M \to F$ whose holonomy diffeomorphisms of the fiber $Y_p = \pi^{-1}(p)$ along the circles $a_1, b_1, \ldots, a_k, b_k$ are equal to $f_1, g_1, \ldots, f_k, g_k$, respectively. Conversely, for any contact structure ξ, transversal to the fibers of the fibration $\pi : M \to F$, the holonomy diffeomorphisms $f_1, g_1, \ldots, f_k, g_k \in \widetilde{\mathrm{Diff}}_+ S^1$ along the circles $a_1, b_1, \ldots, a_k, b_k$ satisfy the inequality (1.4).

REMARK 1.3.12.   **1:** Notice that the commutator $c = [f, g]$ of two diffeomorphisms $f, g : S^1 \to S^1$ admits a canonical covering diffeomorphism $\tilde{c} : \mathbb{R} \to \mathbb{R}$. Indeed, for any choice of diffeomorphisms covering f and g the diffeomorphism $\tilde{c} = [\tilde{f}, \tilde{g}]$ depends only on f, g, and not on the choice of the covering diffeomorphisms $\tilde{f}$ and $\tilde{g}$. Hence the last part of the Proposition 1.3.11 makes sense independent of the choice of the covering diffeomorphisms for the holonomy maps.

2: The non-strict inequality $h(x) \leq x - c$ is the necessary and sufficient condition for existence of a *confoliation* with the prescribed holonomy maps along the circles $a_1, b_1, \ldots, a_k, b_k$.

PROOF. Let $D' \subset D$ be a slightly smaller disc with a smooth boundary $\partial D'$. Let us fix a point $p' \in \partial D'$, which is close to $p \in \partial D$ and choose a trivialization of the bundle F over a neighborhood $U = F \backslash B'$ of the skeleton B. Then for any plane field ξ transversal to the fibers of the fibration $\pi : \pi^{-1}(U) \to U$ the holonomy maps along the circles a_i, b_i, $i = 1 \ldots, k$, can be canonically lifted to the group $\widetilde{\mathrm{Diff}}_+ S^1$. First we define the required ξ over B in such a way that the holonomy maps along the circles a_i, b_i of the bouquet B were equal to the required diffeomorphisms f_i, g_i, $i = 1, \ldots, k$. Next we extend ξ as a contact structure over a neighborhood $U \subset B$. The holonomy diffeomorphism $h' : \tilde{Y}_{p'} \to \tilde{Y}_{p'}$ along the oriented boundary $\partial D'$ is close to h, and thus it satisfies the inequality $h'(x) < x - e$. On the other hand, the trivialization of the bundle π over D' defines another holonomy lifting $h'' : \tilde{Y}_{p'} \to \tilde{Y}_{p'}$ which differs from h' by the translation:

$$h''(x) = h'(x) + e.$$

Thus we get the inequality $h'' < \mathrm{id}$. According to Proposition 1.3.5 there exists a contact structure ξ' over the disc D' whose curvature Ω satisfies the equation

$$\int_{(D',p')} \Omega = h''.$$

The characteristic foliations defined on the boundary $\pi^{-1}(\partial D')$ by the contact structures ξ and ξ' coincide up to a fiber preserving diffeomorphism $\psi : \pi^{-1}(\partial D') \to \pi^{-1}(\partial D')$. This diffeomorphism extends to a contactomorphism between the the germs of the contact structures ξ and ξ' along the surface $\pi^{-1}(\partial D')$ which finishes off the proof of sufficiency of the inequality (1.4).

Conversely, given a contact structure ξ, transversal to the fibration $\pi : M \to F$, its curvature Ω satisfies the inequality

$$\int_{(D',p')} \Omega < \mathrm{id}.$$

On the other hand any trivialization of the bundle π over $U \supset B$ defines a holonomy diffeomorphism $\tilde h' : \tilde Y_{p'} \to \tilde Y_{p'}$ along the circle $\partial D'$ which differs from $\int_{(D',p')} \Omega$ by the translation by $-e$. Hence h' satisfies the inequality $h'(x) < x - e$ for all $x \in \tilde Y_{p'} = \mathbb{R}$. But h' is close to the holonomy map $h : \tilde Y_p \to \tilde Y_p$ along the loop $a_1 b_1 a_1^{-1} b_1^{-1} \ldots a_k b_k a_k^{-1} b_k^{-1}$ which is equal to the product of commutators $[f_1, g_1] \circ \cdots \circ [f_k \circ g_k]$ of holonomy diffeomorphisms along the circles of the bouquet B. Hence, we have the inequality $h(x) < x - e$ as well.   □

Similarly, using 1.3.6 instead of 1.3.5 we can prove

PROPOSITION 1.3.13. *Let* $f_1, g_1, \ldots, f_k, g_k$ *be either*

A): *germs from* G_s, $s = 1, \ldots, \infty$ *of diffeomorphisms* $\mathbb{R}_+ \to \mathbb{R}_+$, *or*
B): *diffeomorphisms of* $\mathbb{R}$ *with compact support.*

Suppose that the product of their commutators satisfies the inequality

(1.5) $$h = [f_1, g_1] \circ \circ \cdots \circ [f_k, g_k] \le \mathrm{id}.$$

Then there exists a positive confoliation ξ *(germ of a confoliation in the case A)) transversal to the fibers of the fibration* $\pi : M \to F$ *whose holonomy diffeomorphisms of the fiber* $Y_p = \pi^{-1}(p)$ *along the circles* $a_1, b_1, \ldots, a_k, b_k$ *are equal, respectively, to* $f_1, g_1, \ldots, f_k, g_k$. *If, in the situation of the case A), we have* $h \in H_l$ *then* ξ *can be chosen in such a way that its curvature* Ω *takes values in* $\mathcal{W}_l$.

REMARK 1.3.14. In case A) the diffeomorphism h cannot belong to H_0, i.e. it must have at least C^1 contact to the identity. Indeed, the commutator $[G_k, G_l] \subset G_{2k+1}$ for $k = 0, \ldots$. In particular, any product of commutators of germs of diffeomorphisms has at least first-order contact to the identity.

Let us return to the case $Y = S^1$.

The next statement is an analog of Milnor-Wood inequality for foliations (see [44], [59]) It was first independently observed by Giroux in [25] and Sato–Tsuboi in [50].

PROPOSITION 1.3.15. *An S^1-bundle $\pi : M \to F$ over an oriented surface F of genus $k > 0$ admits a positive contact structure ξ, transversal to the fibers of the fibration π if and only if the Euler characteristic $\chi(F) = 2 - 2k$ and the Euler number e of the fibration π satisfy the inequality*

$$e \le -\chi(F) = 2k - 2 .$$

PROOF. Everything follows from [59] in combination with Proposition 1.3.11. Indeed, according to [59], for any 1-periodic diffeomorphisms

$$f_1, g_1, \ldots, f_k, g_k : \mathbb{R} \to \mathbb{R}$$

from $\widetilde{\mathrm{Diff}}_+ S^1$ the k-fold product of their commutators

$$h = [\tilde{f}_1, \tilde{g}_1] \circ \cdots \circ [\tilde{f}_k, \tilde{g}_k]$$

satisfies the inequality

$$x - 2k + 1 < h(x) < x + 2k + 1$$

for some $x \in \mathbb{R}$. Hence, Proposition 1.3.11 implies the inequality

$$e \le 2k - 2 = -\chi(F).$$

On the other hand, it is shown in [59] that for any $a < 2k - 1$ there exist diffeomorphisms $f_1, g_1, \ldots, f_k, g_k \in \widetilde{\mathrm{Diff}}_+ S^1$ such that the k-fold product of their commutators $[f_1, g_1] \circ \cdots \circ [f_k, g_k]$ is the translation by $\pm a$. Thus the required contact structure exists in view of 1.3.11.   □

REMARK 1.3.16. Proposition 1.3.15 shows that unlike the case of fibrations over S^2, the trivial S^1-fibration over a surface of genus > 0 admits a contact structure transversal to the fibers. For instance, the form $dz + (\cos z)dx + (\sin z)dy$ defines on $T^3 = T^2 \times S^1$ a contact structure transversal to the fibers. We do not know if a similar phenomenon can appear in higher dimensions. For instance, the 5-dimensional torus T^5 admits a contact structure (see [40]). Does it have a contact structure transversal to the fibers of the trivial fibration $T^5 = T^4 \times S^1 \to T^4$?

1.3.1. Structure of a confoliation near a closed leaf.

As in the previous section we denote by Ξ the space of germs along F of confoliations on $U_+ = F \times \mathbb{R}_+$ which are tangent to F. Any $\xi \in \Xi$ can be viewed as a germ of a $\mathrm{Diff}\,\mathbb{R}_+$-connection on $F \times \mathbb{R}_+$. We consider in this section the problem of classification of germs from Ξ up to a diffeomorphism.

Let us first observe that there are non-diffeomorphic germs. Indeed, let Ω be the curvature form of the connection ξ. The form Ω takes values in the Lie algebra $\mathcal{V}$ of germs at 0 of vector fields on $\mathbb{R}_+$. Set

$$\mathrm{ordcurv}(\xi) = \max\{k \,|\, \Omega \text{ takes values in } \mathcal{V}_k\},$$

where $\mathcal{V}_k \subset \mathcal{V}$ is the subspace of vector fields vanishing at 0 up to order k.

Set

$$\mathrm{ordhol}(\xi) = \max\{k \,|\, \text{ the holonomy along } \Gamma \text{ belongs to } G_k \text{ for all } \Gamma\},$$

where Γ is an oriented curve in F, and G_k consists of germs at 0 of diffeomorhisms which are C^k-tangent to the identity. Thus,

- for any vectors $T, T' \in T_p(F)$, $p \in F$, the vector field $\Omega(T, T')$ can be written as

$$(a_i z^i + \dots)\frac{\partial}{\partial z},$$

where $z \in \mathbb{R}_+, i \geq \operatorname{ordcurv}(\xi)$;
- for any curve Γ the holonomy diffeomorphism $\varphi : \mathbb{R}_+ \to \mathbb{R}_+$ has the form

$$\varphi(z) = z + b_i z^i + \dots, \quad i \geq \operatorname{ordhol}(\xi),$$

if $\operatorname{ordhol}(\xi) > 0$, and the form

$$\varphi(z) = b_1 z + \dots, \quad b_1 > 0,$$

if $\operatorname{ordhol}(\xi) = 0$.

Clearly, $\operatorname{ordcurv}(\xi)$ and $\operatorname{ordhol}(\xi)$ are diffeomorphism invariants of the germ of ξ along F.

According to Proposition 1.3.6 and Remark 1.3.14 we have

(1.6) $\operatorname{ordcurv}(\xi) > 2\operatorname{ordhol}(\xi).$

The inequality (1.6) enables us to define further invariants (compare [57]) of germs of confoliations from Ξ. We will focus here only on the case $\operatorname{ordhol} = 0$. Fix a base point $p \in F$.

Given an oriented curve Γ through the point p set $h_\xi(\Gamma) = \log b_1$, where b_1 is the coefficient of the linear term of the holonomy diffeomorphism of the fiber over the point p along the curve Γ. The proof of the next two lemmas is straightforward.

LEMMA 1.3.17. *Let ξ be a germ from Ξ. Then $h_\xi(\Gamma)$ depends only on the free homotopy class of the loop $\Gamma \in \pi_1(F, p)$. The map $h_\xi : \pi_1(F, p) \to \mathbb{R}$ is a homomorphism into the additive group $\mathbb{R}$, and thus can be viewed as a cohomology class $h_\xi \in H^1(F; \mathbb{R})$.*

The next lemma describes the behavior of the class $h_\xi \in H^1(F, \mathbb{R})$ under a diffeomorphism.

LEMMA 1.3.18. *Let $\xi, \xi' \in \Xi$ and $\operatorname{ordhol}(\xi) = \operatorname{ordhol}(\xi') = 0$. Let $f : F \times \mathbb{R}_+ \to F \times \mathbb{R}_+$ be a diffeomorphism which preserves $F \times 0$ and sends ξ onto ξ'. Then $h_\xi = f^*(h_{\xi'})$.*

Thus the cohomology class $h_\xi \in H^1(F, \mathbb{R})$ provides an obstruction to existence of a diffeomorphism between two germs from Ξ with $\operatorname{ordhol} = 0$.

It is not difficult to show that under a certain non-degeneracy condition the class h_ξ is the only invariant of germs from Ξ with $\operatorname{ordhol} = 0$. Let us denote by Ξ_0 the subset of Ξ which consists of germs ξ such that $\operatorname{ordhol}(\xi) = 0$, and such that the curvature form Ω_ξ pointwise takes values in $\mathcal{W}_1$. Then we have:

PROPOSITION 1.3.19. *Let ξ, ξ' be germs from Ξ_0 with $h_\xi = h_{\xi'}$. Then there exists a diffeomorphism $\Phi : F \times \mathbb{R}_+ \to F \times \mathbb{R}_+$ which preserves $F \times 0$, is isotopic to the identity and sends ξ to ξ'.*

CHAPTER 2

Perturbation of confoliations into contact structures

We will study in this section the possibility of perturbing a foliation (or more generally, a confoliation) $\xi = \{\alpha = 0\}$ into a contact structure. One can try to make the perturbation in three different senses.

We say that ξ can be *linearly deformed* into a positive contact structure if there exists a deformation $\xi_t = \{\alpha_t = 0\}$, $t \in \mathbb{R}_+$, such that $\alpha_0 = \alpha$ and

$$(2.1) \qquad \frac{d(d\alpha_t \wedge \alpha_t)}{dt}\Big|_{t=0} > 0.$$

The inequality (2.1) is equivalent to the inequality

$$(2.2) \qquad \langle \alpha, \beta \rangle \overset{\text{def}}{=} \alpha \wedge d\beta + \beta \wedge d\alpha > 0, \quad \text{where} \quad \beta = \frac{d\alpha_t}{dt}\Big|_{t=0}.$$

It is important to observe that this condition depends on the foliation ξ only and not on the choice of the defining form α. Indeed, we have

$$(2.3) \qquad \langle f\alpha, f\beta \rangle = f^2 \langle \alpha, \beta \rangle$$

for any function f.

Conversely, if there exists a form β which satisfies the inequality (2.2) then the deformation $\alpha_t = \alpha + t\beta$ is the required linear deformation, which defines contact structures $\xi_t = \{\alpha_t = 0\}$ for small $t \neq 0$.

We say that ξ can be *(C^k-)deformed* into a contact structure if there exists a C^k-deformation ξ_t beginning at $\xi_0 = \xi$ such that ξ_t is contact for $t > 0$.

Finally we consider also C^k-*approximations* of ξ by contact structures when it will not be clear that this could be done via a deformation.

2.1. Linear perturbations

Given a foliation $\xi = \{\alpha = 0\}$ we want to find a 1-form β such that $\langle \alpha, \beta \rangle > 0$.

Let us define a real-valued symmetric form $\langle\langle \alpha, \beta \rangle\rangle$ by integrating the 3-form $\langle \alpha, \beta \rangle$:

$$(2.4) \qquad \langle\langle \alpha, \beta \rangle\rangle = \int_M \langle \alpha, \beta \rangle \, .$$

Notice that Stokes' theorem shows that

$$(2.5) \qquad \langle\langle \alpha, \beta \rangle\rangle = -2 \int_M \alpha \wedge d\beta = 2 \int_M \beta \wedge d\alpha \, .$$

PROPOSITION 2.1.1. *If a foliation ξ*

  1. *has a closed leaf with trivial linear holonomy, or*
  2. *can be defined by a closed 1-form α, or even*
  3. *has no holonomy*

then ξ cannot be linearly perturbed into a contact structure.

PROOF. Suppose ξ has a closed leaf F with trivial linear holonomy. Then there exists a trivialization $U = F \times (-\varepsilon, \varepsilon)$ of a tubular neighborhood U of F, such that ξ is transverse to the fibers, and such that parallel transport of normal fibers along any arc preserves the normal coordinate up to first order. Let α be a defining form for ξ normalized to have norm one on the normal fibers. Then $d\alpha = 0$ at the points of F. Hence $\langle \alpha, \beta \rangle = \alpha \wedge d\beta$ along F, and the condition $\langle \alpha, \beta \rangle > 0$ implies that $d\beta$ never vanishes on F, which is impossible since its integral over F is 0. This proves the first part of the proposition.

If α is closed then $\langle\langle \alpha, \beta \rangle\rangle = 0$ for any 1-form β. This proves the second part of the proposition.

Foliations without holonomy are topologically equivalent to foliations given by closed 1-forms (2.1.2), but the equivalence need not be absolutely continuous. Another way to say this is that a foliation without holonomy admits a transverse invariant measure of full support (given by the closed 1-form in the topologically equivalent foliation), but this measure need not be differentiable or even absolutely continuous. (A transverse invariant measure is a measure on the space of leaves in each local coordinate chart, such that the measures in overlapping charts match up.) However, any transverse invariant measure can be approximated by a smooth measure that is almost invariant, *i.e.* if α is a 1-form defining the foliation, there are positive smooth functions f such that the 1-form $f\alpha$ approximates the given transverse invariant measure and $d(f\alpha)$ is uniformly small. [One way to get a suitable smooth approximation is to convolute in the normal direction with a C^∞ bump function of small support.] In light of equations 2.5 and 2.3, we have that for all β,

$$\langle\langle \alpha, \beta \rangle\rangle = 0.$$

$\square$

Here is a statement of the converse:

THEOREM 2.1.2. *Suppose that ξ is a C^2-foliation with holonomy, and each of its closed leaves has a curve with non-trivial linear holonomy. Then ξ can be linearly deformed into a contact structure.*

The proof of Theorem 2.1.2 is given in Section 2.8 below.

2.2. Conformally-Anosov Flows

The content of this section has a large overlap with a recent paper of Y. Mitsumatsu (see [**46**]).

As we had seen above, the existence of a linear deformation of a foliation $\xi = \{\alpha = 0\}$ into a contact structure amounts to the existence of a 1-form β which satisfies the inequality (2.2). Let us consider when the form β itself can be chosen to define a foliation.

PROPOSITION 2.2.1. *Suppose that 1-forms α and β define foliations ξ and η, and satisfy the inequality (2.2). Then ξ and η are transversal, and for all $t \in (0, \pi)$*

different from $\pi/2$ the form $(\cos t)\alpha + (\sin t)\beta$ defines a contact structure, which is positive when $t \in (0, \pi/2)$, and negative when $t \in (\pi/2, \pi)$.

PROOF. At each point $p \in M$ at least one of the forms $d\alpha|_\eta$ or $d\beta|_\xi$ does not vanish. On the other hand $d\alpha|_\xi = 0$. Thus ξ and η are pointwise different, and hence transversal. The second part follows from the identity

$$((\cos t)\alpha + (\sin t)\beta) \wedge d\,((\cos t)\alpha + (\sin t)\beta) = \frac{1}{2}(\sin 2t)\langle \alpha, \beta \rangle.$$

$\square$

It turns out that the appearance of two 1-forms as in Proposition 2.2.1 has an important dynamical meaning which we discuss below. Let us begin with the following simple

LEMMA 2.2.2. *Suppose a C^1-smooth plane field ξ admits a tangent vector field X whose flow preserves ξ. Then ξ is a foliation.*

PROOF. If $\xi = \{\alpha = 0\}$ then $L_X\alpha = h\alpha$ for a function $h : M \to \mathbb{R}$. But $L_X\alpha = X \rfloor d\alpha$. Hence,

$$X \rfloor (\alpha \wedge d\alpha) = \alpha(X)d\alpha - \alpha \wedge (X \rfloor d\alpha) = -\alpha \wedge L_X\alpha = -h\alpha \wedge \alpha = 0.$$

Therefore, $\alpha \wedge d\alpha = 0$, and ξ is a foliation. $\square$

Let us recall that a flow $X^t : M \to M$ generated by a vector field X is called *Anosov* if there exists a continuous Riemannian metric on M and a continuous, splitting $TM = N_+ \oplus N_- \oplus \{\lambda X\}$ of the tangent bundle TM into the direct sum of 1-dimensional bundles, such that the splitting is invariant under the flow and the differential $dX^t : T(M) \to T(M)$ acts by dialations on N_+ and by contractions on N_-. In other words, there exists a constant $C > 0$ such that for all $t > 0$ we have

$$||dX^t(v)|| \geq e^{Ct}||v||$$

for $v \in N_+$ and

$$||dX^t(v)|| \leq e^{-Ct}||v||$$

for $v \in N_-$.

If the invariant plane fields ξ_+ and ξ_- generated by (X, N_+) and (X, N_-), respectively, are C^1-smooth, then according to Lemma 2.2.2 they are integrable. However, in many geometrically interesting cases, the plane fields ξ_+ and ξ_- are not smooth. Nevertheless dynamical arguments (see [4]) show that even in the non-smooth case ξ_+ and ξ_- are tangent to invariant C^0-foliations, which are called the *unstable and stable foliations* of the Anosov flow X^t.

Generalizing the notion of Anosov flow, we will call a flow X^t *conformally-Anosov* (Mitsumatsu in [46] called it *projectively Anosov*) if there exists a continuous Riemannian metric and a continuous, invariant splitting $N = N_+ \oplus N_-$, as in the definition of Anosov flow, such that the differential $dX^t : T(M) \to T(M)$ acts by dialations on N_+ and by contractions on N_- *after $dX^t|_N$ has been renormalized to have determinant 1*. In other words, there exists a constant $C > 0$ such that for all $t > 0$ the following inequality

(2.6) $$||dX^t(v_+)||/||dX^t(v_-)|| \geq e^{Ct}||v_+||/||v_-||$$

holds for any two non-zero vectors $v_+ \in N_+, v_- \in N_-$.

As in the Anosov case, if the invariant plane fields ξ_+ and ξ_- are smooth then they are integrable. However, unlike the Anosov case, ξ_+ and ξ_- need not to be integrable if they are not smooth, as an example 2.2.9 below shows.

PROPOSITION 2.2.3. *Suppose X^t is a conformally-Anosov flow with C^1-smooth stable and unstable foliations ξ_- and ξ_+. Then ξ_- and ξ_+ can be defined by 1-forms α_- and α_+ such that*

$$(2.7) \qquad\qquad\qquad \langle \alpha_+, \alpha_- \rangle > 0.$$

Conversely, suppose forms α_+ and α_- satisfy the inequality (2.7) and define the foliations ξ_+ and ξ_-, then any integrable non-vanishing vector field $X \in \xi_+ \cap \xi_-$ defines a conformally-Anosov flow (ξ_+ and ξ_- are transversal according to 2.2.1) .

PROOF. Let us normalize the defining forms $\alpha_\pm$ for the foliations $\xi_\pm$, so that $\|\alpha_\pm\| = 1$ in terms of the Riemannian metric implied by the definition of a conformally-Anosov flow. We have $L_X \alpha_\pm = f_\pm \alpha_\pm$, and the inequality 2.7 is equivalent to the condition

$$(2.8) \qquad\qquad\qquad f_+ - f_- > 0.$$

On the other hand,

$$(2.9) \qquad\qquad X \rfloor \langle \alpha_+, \alpha_- \rangle = (f_+ - f_-)\alpha_+ \wedge \alpha_- .$$

Therefore, if the inequality (2.8) is satisfied then so is the inequality (2.7). Conversely, suppose (2.7) holds. Then ξ_+ and ξ_- are transversal. Therefore $\alpha_+ \wedge \alpha_- \neq 0$, and (2.9) implies the inequality (2.8).   □

Notice that the metric adapted to a conformally-Anosov flow can always be chosen in such a way that the invariant plane fields ξ_+ and ξ_- are orthogonal. In this case the contact structure defined by the form $(\cos t)\alpha_+ + (\sin t)\alpha_-$, $t \in (0, \pi)$, $t \neq \pi/2$, contains the vector field generating the flow and makes a constant angle t with the planes ξ_+ and ξ_-. The next proposition shows that even in the case when the stable and unstable plane fields ξ_+ and ξ_- are not smooth, the above plane fields $(\cos t)\alpha_+ + (\sin t)\alpha_-$, $t \in (0, \pi)$, $t \neq \pi/2$, can be perturbed into smooth contact structures.

PROPOSITION 2.2.4. *Suppose X^t is a conformally-Anosov flow with stable and unstable plane fields $\xi_- = \{\alpha_- = 0\}$ and $\xi_+ = \{\alpha_+ = 0\}$ which are not assumed to be smooth. Let $\tilde\alpha_+$ and $\tilde\alpha_-$ be smooth approximations to α_+ and α_- which annihilate the vector field X. Then for any $\varepsilon > 0$ there exists $T > 0$ such that the plane field*

$$\tilde\xi_t = \{(\cos t)(X^T)^* \alpha_+ + (\sin t)(X^{-T})^* \alpha_- = 0\}$$

is a positive contact structure for any $t \in (\varepsilon, \pi/2 - \varepsilon)$, and negative contact structure for any $t \in (\pi/2 + \varepsilon, \pi - \varepsilon)$.

PROOF. Even in the non-smooth case the forms α_+ and α_- can be chosen C^∞-smooth along the flow-lines. Thus the forms $(X^{\pm T})\tilde\alpha_\pm$ C^∞-converge to $\alpha_\pm$ when $T \to \infty$ along the flow lines. Thus when T is sufficiently large then the criterion 1.1.6 applied to the plane field $\tilde\xi_t$ guarantees that it is a positive contact structure for $t \in (\varepsilon, \pi/2 - \varepsilon)$ and negative contact structure when $t \in (\pi/2 + \varepsilon, \pi - \varepsilon)$.   □

Thus there exists a family $\xi_t^+, t \in (0,1)$, of smooth positive contact structures and a family $\xi_t^-, t \in (0,1)$, of smooth negative contact structures which continuously converge when $t \to 0$ and $t \to 1$ to the invariant plane fields ξ_+ and ξ_-.

Conversely, a conformally-Anosov flow can be characterized, as the next proposition shows, by the existence of two transversal contact structures ξ_+ and ξ_-, the first one positive and the second negative.

REMARK 2.2.5. Mitsumatsu in [46] calls such pair (ξ_+, ξ_-) a *bi-contact structure*. He also proves a proposition similar to our propositions 2.2.4 and 2.2.6

PROPOSITION 2.2.6. *Suppose ξ_+ and ξ_- are two transversal contact structures, the first one positive and the second negative. Then the vector field directing the intersection $\xi_+ \cap \xi_-$ defines a conformally-Anosov flow.*

PROOF. Let us choose a plane field ν transversal to X and orient ν in such a way that together with X it defines the given orientation of M. Set $\lambda_\pm = \nu \cap \xi_\pm$. As in 1.1.6 denote by $\lambda_\pm^t(p), t \in \mathbb{R}$, the line

$$d_{X^{-t}(p)}X^t(\xi_\pm(p)) \cap \nu,$$

where $d_{X^{-t}(p)}X^t : T_{X^{-t}(p)}(M) \to T_p(M)$ is the differential of the map X^t at the point $X^{-t}(p)$. Denote by $\theta_\pm^t(p)$ the angle between the lines $\lambda_\pm^0(p)$ and $\lambda_\pm^t(p)$. According to 1.1.6 we have

$$\frac{d\theta_+^t(p)}{dt} > 0 \quad \text{and} \quad \frac{d\theta_-^t(p)}{dt} < 0$$

for all $p \in M, t \in \mathbb{R}$. Notice that the sum $\theta_+^t(p) + \theta_-^t(p)$ is bounded above by the angle between the lines $\lambda_+^0(p)$ and $\lambda_-^0(p)$. Thus the families of lines $\lambda_\pm^t(p)$ converge when $t \to +\infty$ to the lines $\lambda_\pm^\infty(p) \subset \nu(p)$. Both line fields $\lambda_\pm^\infty$ are continuous and invariant under the flow $X^t, t \in \mathbb{R}$. We claim that $\lambda_+^\infty = \lambda_-^\infty$. Indeed, suppose that $\lambda_+^\infty \neq \lambda_-^\infty$. If there exists a periodic point p of the flow, *i.e.* $X^T(p) = p$ for $T > 0$, then the linear map $D_p X^T : \nu(p) \to \nu(p)$ leaves invariant two distinct lines $\lambda_+^\infty(p), \lambda_-^\infty(p) \subset \nu(p)$ while it turns the lines $\lambda_\pm(p)$ towards each other. This is impossible because $\lambda_\pm$ do not separate $\lambda_+^\infty(p)$ and $\lambda_-^\infty(p)$. If there are no periodic points then a similar argument works for an accumulation point of any orbit of the flow X^t. Thus, $\lambda_+^\infty = \lambda_-^\infty = \lambda^\infty$. Similarly we define the invariant line field $\lambda^{-\infty}$ taking the limits $\lambda_\pm^t(p)$ for $t \to -\infty$.

Take now a continuous Riemannian metric on the manifold as follows. We declare X, λ^∞ and $\lambda^{-\infty}$ to be orthogonal, choose the vector field X to be of length 1, define the metric arbitrarily on λ^∞, and choose the scale on $\lambda^{-\infty}$ in such a way that λ_+ and λ^∞ form a constant angle equal $\pi/4$. Then the Proposition 1.1.6 guarantees that the flow X^t is conformally-Anosov. □

Examples of conformally-Anosov flows on T^3. Below we give examples of conformally-Anosov flows on T^3 with smooth and non-smooth invariant plane fields. The non-smooth example 2.2.9 shows that unlike the Anosov case the invariant plane fields need not to be integrable.

EXAMPLE 2.2.7. Let (x, y, z) be coordinates in $T^3 = \mathbb{R}^3/\mathbb{Z}^3$. Take two 1-forms $\alpha = dz + \cos z dx$ and $\beta = dz + \sin z dy$. Then $\langle \alpha, \beta \rangle > 0$ and hence the foliations $\{\alpha = 0\}$ and $\{\beta = 0\}$ serve as the invariant foliations of a conformally-Anosov flow.

We generalize the previous example in the following

PROPOSITION 2.2.8. *Let η be a 1-dimensional foliation on the torus T^2 which has closed leaves and which has non-trivial linear holonomy along each closed leaf. Then the 2-dimensional product-foliation $\tilde{\eta}$ on $T^3 = T^2 \times S^1$ whose leaves are products of the leaves of η by S^1 serves as an invariant foliation of a conformally-Anosov flow.*

PROOF. There exists a splitting $T^2 = S^1 \times S^1$, such that in the coordinates (x, y) corresponding to the splitting the foliation η can be defined by the 1-form

$$\alpha = f(x)dx + g(x, y)dy,$$

such that along any horizontal curve $y = C$, $g(x, y)$ looks roughly like $\sin(x\pi/2)$, that is,

- the 0 set of g consists of the circles

$$\{x = \frac{k}{n}\}, \quad k = 0, \ldots, n - 1 :$$

- the 0-set of the derivative $\frac{\partial g(x,y)}{\partial x}$ consists of the circles

$$\{x = \frac{2k + 1}{2n}\}, \quad k = 0, \ldots, n - 1.$$

In particular, if $g(x, y) = 0$ then $\frac{\partial g(x,y)}{\partial x} \neq 0$.

Then, there exists a function $h : S^1 \to \mathbb{R}$ such that along any horizontal circle $y = C$ the pair (g, h) progresses counterclockwise about the origin, that is,

$$h(x)\frac{\partial g(x, y)}{\partial x} - h'(x)g(x, y) > 0$$

for all $x, y \in S^1$.

Set $\beta = dx + h(x)dz$. Then we have

$$\langle \alpha, \beta \rangle = (h(x)\frac{\partial g(x, y)}{\partial x} - h'(x)g(x, y))dx \wedge dy \wedge dz > 0,$$

and thus the foliations $\eta = \{\alpha = 0\}$ and $\xi = \{\beta = 0\}$ are transversal and serve as the stable and unstable foliations of a conformally-Anosov flow. □

Thus, T^3 has many conformally-Anosov flows while it has no Anosov flows. See Mitsumatsu's paper [**46**] for more examples of conformally-Anosov, but not Anosov, flows. He calls such flows *essentially projectively Anosov*.

EXAMPLE 2.2.9. (*Conformally-Anosov flow with no invariant foliations*) We describe here an example of a smooth, conformally-Anosov diffeomorphism of T^2 (whose suspension is a conformally-Anosov flow on T^3) such that there is no invariant foliation, but a kind of branching foliation. It is not clear how much more pathological the invariant structure can get, but at least this example shows that the invariant foliation definition diverges from the infinitesimal definition, and that one cannot automatically get an invariant foliation of a conformally-Anosov flow from a positive and a negative contact structures which are transverse to each other.

Start with an example of a conformally-Anosov diffeomorphism of T^2 that has a circle of fixed points, with a attracting neighborhood. The examples described in 2.2.7 and 2.2.8 are suspensions of diffeomorphisms like this, but perhaps the best construction is to start with the 2-torus T^2 considered as the factor space of $R^2 \setminus \mathbf{0}$

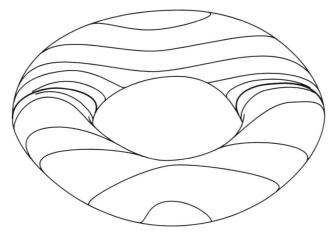

FIGURE 2.1. One of the two foliations with two Reeb components arising on T^2. The two closed leaves appear as horizontal circles in this picture.

by the subgroup generated by the homothety $v \mapsto 2v$, $v \in \mathbb{R}^2$, and with the two foliations coming from the lines parallel to the x and y axes. On the torus, these give rise to foliations with two Reeb components (see 2.1).

These foliations are invariant by the map

$$(x, y) \mapsto (2x, \frac{y}{2})$$

which distorts the conformal structure in the right manner to be conformally-Anosov. The x-axis and the y-axis each map to two circles of fixed points, and the positive x-axis is a good circle C to start with in this construction.[1]

Now modify the diffeomorphism near C to get a new diffeomorphism φ such that (see Figure 2.2):

- C is still invariant;
- the stable foliation (normal to C) is unchanged, and so it remains invariant under φ;
- the dynamics on C has one linearly attracting fixed point A and one linearly repelling fixed point R;
- near the attracting fixed point on C, φ contracts at a faster linear rate in the normal direction than in the tangent direction;
- on the stable leaf through R, the dynamics are that R repels at a slower linear rate than in the direction of C, and R is bracketed by two attracting fixed points H_+ and H_- (which repel in the normal direction);
- all the stable leaves that intersect C are attracting toward C outside of some compact set, and in fact, the dynamics is unmodified outside some compact subset of the union of stable leaves that intersect C.

Now what happens is that any orbit not on the stable leaf of A or R tends toward A, with tangential approach along some power curve (the power depending on the two eigenvalues at A). The unstable manifolds through H_+ and H_- are easy to define dynamically near those two points. They end up both heading toward A,

---

[1] A similar example is described in [**19**].

FIGURE 2.2. Dynamics of φ near the curve C.

FIGURE 2.3. The invariant foliation before the perturbation.

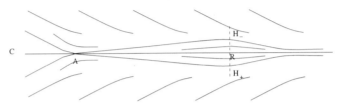

FIGURE 2.4. The "delaminated" invariant foliation after the perturbation.

with equal tangents. They enclose a whole bundle of stable manifolds that behave similarly.

The diffeomorphism φ is still conformally-Anosov because it is (or can be made so) near the stable leaf of R, and because all other orbits outside a neighborhood of R tend in a bounded number of steps to a small linearizable neighborhood of A, where the linear effect at A has all the time it wants to over-ride any contrary tendency on the journey there.

In the global picture, one of the two closed leaves of one of the two unstable foliations has become "delaminated", with part of it still pinched to a point, but part of it flared out to enclose an open set. See Figure 2.3 and Figure 2.4.

Suspending φ, we get a conformally-Anosov flow on T^3 which has no invariant foliations.

2.3. Non-linear deformations

We give here an example of a non-linear deformation of a foliation into a contact structure in the situation when the linear deformation does not exist.

Let ξ_0 be the foliation of the torus T^3 by the 2-dimensional tori $T^2 \times p$, $p \in S^1$. If $x, y, \theta \in [0, 2\pi)$ are coordinates on T^3, then ξ_0 is given by the Pfaffian equation $d\theta = 0$. It is straightforward to check that

PROPOSITION 2.3.1. *For any integer $n > 0$, and any $t > 0$ the form*

$$\alpha_n^t = d\theta + t(\cos n\theta dx + \sin n\theta dy)$$

defines a contact structures on T^3.

According to a theorem, independently proven by Giroux (see [**24**]) and Kanda (see [**37**]), the contact structures $\xi_n = \{\alpha_n^1 = 0\}$ are pairwise non-diffeomorphic. On the other hand, according to Gray's theorem [**29**] the contact structures $\{\alpha_n^t = 0\}$ are isotopic, when n is fixed and t varies in $(0, \infty)$. Thus

PROPOSITION 2.3.2. *The foliation ξ_0 can be C^∞-deformed into an infinite number of non-equivalent contact structures.*

REMARK 2.3.3. It is likely that any foliation on an orientable manifold can be deformed into a contact structure. However, we are able to prove only a weaker approximation result (see Theorem 2.4.1 which we consider in the next section, and Proposition 2.9.2 below).

2.4. Approximations of foliations by contact structures

According to Corollary 1.3.8 the foliation ζ on $S^2 \times S^1$ cannot be C^0 approximated by contact structures (and even by somewhere positive confoliations). Moreover, Proposition 2.1.1 shows that in certain cases a foliation cannot be *linearly deformed* into a contact structure. Despite these negative facts we have

THEOREM 2.4.1. *Suppose that a C^2-confoliation ξ on an oriented 3-manifold is different from the foliation ζ on $S^2 \times S^1$. Then ξ can be C^0-approximated by contact structures. When ξ is a foliation it can be approximated both by positive and negative contact structures.*

The proof of the theorem is given in sections 2.5–2.8 below. We will consider only the case when M is closed and ξ is co-orientable. The necessary adjustments for the non-co-orientable case are described in Section 2.9.1 below. We leave to the reader the minor changes necessary for the non-compact case.

2.5. Perturbation near holonomy curves

Let Γ be an embedded closed curve in a leaf S of a co-oriented foliation ξ. By a *positive (resp. negative) semi–neighborhood* of Γ we mean the positive (resp. negative) component of $U \setminus S$ for a neighborhood U of Γ. There are a number of situations when it is possible to make a small perturbation of the foliation to be contact near Γ. In this section, we prove the following:

PROPOSITION 2.5.1. *Let (M, ξ) be a C^k-foliation, $k \geq 0$.*

a): *Suppose Γ is a curve with non-trivial linear holonomy. Then ξ can be C^k-perturbed into a positive or negative confoliation (at our choice) which is contact in a neighborhood U of Γ, coincides with ξ outside of a bigger neighborhood, and is diffeomorphic to ξ outside of U, see Figure 2.5.*

b): *Suppose the curve Γ has attracting holonomy on its positive side. Then there exists a positive (or negative) confoliation $\tilde{\xi}$ which is contact in a positive semi-neighborhood U_+ of Γ, C^0-close to ξ, diffeomorphic to ξ outside U_+, and coincides with ξ outside a bigger positive semi-neighborhood.*

c): *Suppose Γ has sometimes attracting (two-sided) holonomy. Then one can C^0-perturb ξ into a positive (or negative) confoliation which is contact in a neighborhood U of Γ, diffeomorphic to ξ outside of U, and coincides with ξ outside a bigger neighborhood. See Figure 2.6.*

To prove Proposition 2.5.1 we will need several lemmas.

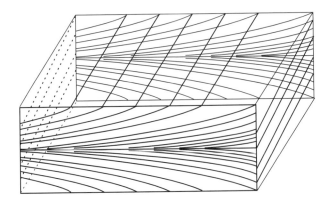

FIGURE 2.5.   Local perturbation to contact near a curve with repelling holonomy, as in proposition 2.5.1 a). This figure represents a solid torus around Γ. The left and right rectangles are cross-sections of the solid torus, and are identified. The front and back faces of the figure describe the holonomy f around Γ, which is repelling left to right. The cross-section is the mapping cylinder of a diffeomorphism g of the interval, such that for all x in the open interval $g^{-1}f^{-1}gf(x) < x$, *i.e.*, when you follow the leaves once counterclockwise around the box, whenever this can be defined you descend. By 1.3.5, the solid torus fills in with a contact structure.

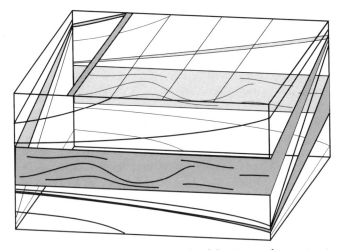

FIGURE 2.6.   Local perturbation of a foliation to be contact near an annulus such that the foliation "expands" across its boundary. Note how each leaf when followed counterclockwise around the boundary of the box spirals downward (*cf.* proposition 2.5.1 c).)

LEMMA 2.5.2. *Let v_x be a compact family of smooth functions on $[-1,1]$ such that $v_x(0) = 0$, and v_x is monotonically increasing for all x. Then there exists a C^∞-close to the identity diffeomorphism $f : [-1,1] \to [-1,1]$ which is C^∞-tangent to the identity near the ends and such that*

(2.10) $f'(z)v_x(z) > v_x(f(z))$ *for all* $z \in (-1,1)$ *and all* x.

Similarly, one can find a diffeomorphism f for which the opposite inequality holds:

(2.11) $f'(z)v_x(z) < v_x(f(z))$ *for all* $z \in (-1,1)$ *and all* x.

PROOF. Choose a sufficiently small $\varepsilon > 0$. The inequality (2.10) will be obviously satisfied by choosing f to be any diffeomorphism $[-1,1] \to [-1,1]$ which is C^∞ tangent to the identity at the ends and such that $f'(x) > 1$ for $x \in (1-\varepsilon, 1)$, $f'(x) = 1$ for $x \in [-1+\varepsilon, 1-\varepsilon]$ and $f'(x) < 1$ for $x \in (-1, -1+\varepsilon)$. Similarly, to satisfy the second inequality (2.11) one need to choose f to be attracting on $[1-\varepsilon, 1]$, dilating on $[-1, -1+\varepsilon]$, and to be equal to a translation on $[-1+\varepsilon, 1-\varepsilon]$.   □

LEMMA 2.5.3. *Let v_x be a compact family of C^1-functions on $(-1,1)$ such that $v_x(0) = 0$ and there exist sequences $z_n \downarrow +0$ and $z'_n \uparrow -0$ such that $v_x(z_n) > 0$ and $v_x(z'_n) < 0$ for all $n = 1,\dots$ and all x. Then for any sufficiently small $\varepsilon > 0$ there exists a diffeomorphism $f : (-1,1) \to (-1,1)$ which is fixed outside the interval $(-\varepsilon, \varepsilon)$ and such that*

(2.12) $f'(z)v_x(z) > v_x(f(z))$ *for all* $z \in (-\varepsilon, \varepsilon)$ *and all* x.

One can also find a diffeomorphism f with similar properties which satisfies the opposite inequality

(2.13) $f'(z)v_x(z) < v_x(f(z))$ *for all* $z \in (-\varepsilon, \varepsilon)$ *and all* x.

The proof is similar to 2.5.2 and we leave it to the reader.

PROOF OF PROPOSITION 2.5.1. Let S be the leaf which contains the curve Γ. Let us split a neighborhood U of Γ into the product $U = \Gamma \times [-1,1] \times [-1,1] = \Gamma \times D_1 \times D_2$ in such a way that

- $\Gamma \times 0 \times 0 = \Gamma$;
- $\Gamma \times D_1 \times 0 \subset S$;
- $U_+ = \Gamma \times D_1 \times (0,1)$ is a positive semi-neighborhood of Γ;
- the intervals $x \times D_1 \times z$, $x \in \Gamma, z \in D_2$, are tangent to ξ.

Let (x, y, z) be the coordinates in U corresponding to the above splitting, so that x is the cyclic coordinate along Γ, y corresponds to the factor D_1, and z corresponds to the factor D_2, respectively. In these coordinates the foliation $\xi|_U$ can be defined by the 1-form $\alpha = dz + a(x,z)dx$.

Let us consider first the cases a) and c). Set $v_x(z) = a(x,z)$, $x \in \Gamma$, $z \in [-1,1]$. If the neighborhood U is chosen sufficiently small then the family $\{v_x\}_{x\in\Gamma}$ satisfies the conditions of Lemma 2.5.2 in the case a) and the conditions of Lemma 2.5.3 in the case c). Let $f : [-1,1] \to [-1,1]$ be the diffeomorphism provided by these lemmas, which satisfies the second of the two inequalities. Thus we have the inequality

(2.14) $f'(z)v_x(z) < v_x(f(z))$,

FIGURE 2.7. The characteristic foliation on the annulus $\Gamma \times 0 \times D_2$ and its image under the diffeomorphism $(x, y, z) \mapsto (x, y, f(z))$. Note how the tangent lines in the image are rotated clockwise from the original.

which holds for all $x \in \Gamma$ and all $z \in (-1, 1)$ in the case a), and for all z sufficiently close to 0 in the case c). In the first case f can be chosen C^∞-close to the identity, while in the second case one can achieve only C^0-closeness.

Consider an isotopy $f_t : [-1, 1] \to [-1, 1]$, $t \in [-1, -1/2]$, which connects the identity map f_{-1} with the diffeomorphism $f_{-1/2} = f$, and such that the isotopy is tangent to the identity at the end points of $[-1, 1]$ and is fixed for t close to -1 and $-1/2$. Define a diffeomorphism $F : U^- = U \cap \{y \le -1/2\} \to U^-$ by the formula

$$F(x, y, z) = (x, y, f_y(z)), \ x \in \Gamma, \ y \in [-1, -1/2], \ z \in [-1, 1]$$

and consider the foliation $\tilde{\xi} = F_*(\xi)$ defined by the form $\tilde{\alpha} = (F^{-1})^*(\alpha)$, see Figure 2.7.

For y close to $-1/2$ we have

$$\tilde{\alpha} = g'(z)dz + v_x(g(z))dx,$$

where $g = f^{-1}$. Set $g(z) = u$ and notice that

$$g'(z) = \frac{1}{f'(u)}.$$

Thus $\tilde{\xi}$ can be defined for y near $-1/2$ by the form

$$\hat{\alpha} = dz + v_x(u)f'(u)dx.$$

Observe that in view of the inequality (2.14) we have

$$v_x(u)f'(u) < v_x(z).$$

Take now a monotone surjective function $\theta : [-1/2, 1] \to I$ such that $\theta'(y) > 0$ for all $y \in (-1/2, 1)$, and is C^∞-flat at the end-points. Extend $\hat{\alpha}$ to the whole neighborhood U by setting

$$\hat{\alpha} = \theta(y)\alpha + (1 - \theta(y)\hat{\alpha}) = dz + (v_x(z) + (1 - \theta(y))(v_x(u)f'(u) - v_x(z)))\, dx$$
$$= dz + \tilde{v}_x(y, z)dx,$$

where

$$\frac{\partial \tilde{v}_x}{\partial y} < 0$$

in $U' = \Gamma \times (-1/2, 1) \times (-1, 1)$ in the case a) and inside a smaller neighborhood $U'' = U' \cap \{z \le \varepsilon\}$ in the case c). Thus using Criterion 1.1.5 we see that the distribution $\hat{\xi}$ defined in U by the form $\hat{\alpha}$ is a confoliation which is a positive contact structure in U' or U''. Moreover it C^k-agrees with the foliation ξ at the boundary of the semi-neighborhood U_+, and is diffeomorphic to ξ outside the neighborhood where it is contact. The confoliations ξ and $\hat{\xi}$ are C^k-close in the case a), and

C^0-close in the case c). Similarly, if we need to get a negative confoliation then we should use the diffeomorphism f which satisfies the inequality (2.10) instead of (2.11). Its existence is guaranteed by the same Lemma 2.5.2.

Let us consider now the case b). Set $\overline{U}_+ = U \cap \{z \geq 0\}$, $U_{+,\delta} = U \cap \{0 \leq z < \delta\}$ for a positive $\delta < 1$. In the coordinates (x, y, z) corresponding to the splitting we have

$$\xi|_{\overline{U}_+} = \{\alpha = 0\} \quad \text{where} \quad \alpha = dz + v(x, z)dx = 0.$$

The function $v : \overline{U}_+ \to \mathbb{R}$ is positive for $z > 0$, and vanishes when $z = 0$. Given an $\varepsilon > 0$ one has $|v|_{U_{+,\delta}}| < \varepsilon$ for a sufficiently small δ. Moreover, changing if necessary the splitting of the neighborhood U one can arrange that the function v is constant for $\delta/2 \leq |z| \leq \delta$. Set

$$V(z) = \frac{\max\limits_{x \in \Gamma} v(x, z)}{\min\limits_{x \in \Gamma} v(x, z)} \quad z \in (0, \delta].$$

Take now a strictly increasing function $\lambda : (0, \delta] \to \mathbb{R}$ which vanishes with all its derivatives at the point δ and such that

(2.15) $\lambda'(z) > V(z)$

for all $z \in (0, \delta/2]$. Consider a diffeomorphism $f : \Gamma \times (0, \delta] \to \Gamma \times (0, \delta]$ given by the formula

$$f(x, z) = (x - \lambda(z), z).$$

Let $\mu : [1/2, 1] \to [1/2, 1]$ be a smooth monotone function which is equal to 1 near $1/2$, and equal to 0 near 1. Then $f_t(x, z) = (x - \mu(t)\lambda(z), z)$, $t \in [1/2, 1]$, is an isotopy connecting $f_{1/2} = f$ with $f_1 = \text{id}$. Define a diffeomorphism $F : U_{+,\delta} \cap \{-1 < y < -1/2\} \to U_{+,\delta} \cap \{-1 < y < -1/2\}$ by the formula

$$F(x, y, z) = (x, y, f_{|y|}(x, z)), \quad x \in \Gamma, \ y \in [-1, -1/2], \ z \in [0, \delta).$$

Then the foliation $\tilde{\xi} = F_*(\xi)$ on $U_{+,\delta} \cap \{-1 < y < -1/2\}$ is defined by the form $\tilde{\alpha} = (F^{-1})^*(\alpha)$.

Notice that if ξ is C^1-smooth then it can be easily arranged that the function V is bounded. In this case the function λ can be chosen smooth at the point 0 and thus the diffeomorphism f defined on the closure $\Gamma \times [0, \delta]$. However, even when ξ is not smooth the foliation $\tilde{\xi}$ is continuous up to the boundary.

Near $y = -1/2$ we can write

$$\tilde{\xi} = \{dz + \frac{\tilde{v}(x, z)}{1 + \tilde{v}(x, z)\lambda'(z)} dx\},$$

where $\tilde{v}(x, z) = v(x + \lambda(z), z)$. Notice that

(2.16) $\frac{\tilde{v}(x, z)}{1 + \tilde{v}(x, z)\lambda'(z)} < v(x, z)$

for all $z \in (0, \delta)$. Indeed, for $z \geq \delta/2$ this follows from the fact that v is constant and therefore $\tilde{v}(x, z) = v(x, z)$. For $z \in (0, \delta/2)$ the inequality (2.16) follows from the inequality (2.15).

The rest of the proof is similar to the cases a) and c). Take a monotone function $\theta : [-1/2, 1] \to I$ such that $\theta'(y) > 0$ for all $y \in (-1/2, 1)$, $\theta(-1/2) = 0$, $\theta(1) = 1$, and all derivatives of θ vanish at the points $-1/2$ and 1. Extend the plane field $\tilde{\xi}$ from $U_{+,\delta} \cap \{-1 < y < -1/2\}$ to $U_{+,\delta}$ setting

$$\tilde{\xi} = dz + (\theta(y)v(x,z) + (1 - \theta(y))\hat{v})\,dx,$$

where

$$\hat{v}(x,z) = \frac{\tilde{v}(x,z)}{1 + \tilde{v}(x,z)\lambda'(z)}.$$

Using the criterion 1.1.5 we see that the distribution $\tilde{\xi}$ is a confoliation which is a negative contact structure in $U_{+,\delta} \cap \{-1/2 < y < 1\}$. Moreover, it is ε-close to the foliation ξ in C^0-norm, agrees with this foliation at the boundary of the semi-neighborhood U_+, and is diffeomorphic to ξ outside the neighborhood $U_{+,\delta} \cap \{-1/2 < y < 1\}$. The perturbation of ξ into a positive confoliation is similar and left to the reader.   □

2.6. Alternative approaches to Proposition 2.5.1

We sketch in this section other ways to prove some parts of the Proposition 2.5.1 which, although essentially equivalent, help to clarify the construction.

We begin with another proof of Proposition 2.5.1a) which clarifies the linear character of the perturbation in this case.

PROPOSITION 2.6.1. *Let* $\xi = \{\alpha = 0\}$ *be a foliation which admits a curve* Γ *with non-trivial linear holonomy. Then* ξ *can be linearly deformed into a confoliation, contact in a neighborhood of* Γ*. In other words, there exists a form* β *on a neighborhood* $U \supset \Gamma$ *which vanishes outside* U *and satisfies the inequality*

$$\langle \alpha, \beta \rangle > 0$$

inside U*. Moreover, one can choose* β *in such a way that* $\beta|_U$ *defines a foliation, i.e the pair of* (α, β) *defines a conformally-Anosov flow on* U*.*

PROOF. The foliation ξ can be defined in a neighborhood $U = \Gamma \times D_1 \times D_2$ by the form $\alpha = dz + v(x,z)dx$, where $\frac{\partial v}{\partial z} \geq C$ for a positive constant $C > 0$. Take a monotone smooth function $h : \mathbb{R} \to [0,1]$ which is equal to 1 near 0, positive on $[0,1)$, and vanishes on $[1,\infty)$. It is straightforward to check that the form $\beta = h(y^2 + z^2)dy$ satisfies the inequality

$$\langle \alpha, \beta \rangle > 0$$

everywhere inside U.   □

Next we consider a version of 2.5.1c) when one can construct a C^k-deformation instead of C^0-perturbation.

A germ at 0 of a diffeomorphism f is called *nice* if the germ $f(x) - x$ is a germ of a (not necessarily strictly) monotone function. Clearly, any f which has finite order of tangency to the identity is nice. But a generic C^∞-flat germ cannot be always made nice even by conjugating it by a germ of a C^1-smooth diffeomorphism. Analyzing the proof of 2.5.1c) we can get the following stronger version for the case when the holonomy map is nice.

PROPOSITION 2.6.2. *Suppose* Γ *is a curve with nice attracting (two-sided) holonomy of a* C^k*-foliation* ξ*,* $k = 1, \dots, \infty$*. Then one can* C^k*-deform* ξ *into a positive (or negative) confoliation which is contact in a neighborhood* U *of* Γ*, diffeomorphic to* ξ *outside of* U*, and coincides with* ξ *outside a bigger neighborhood.*

2.7. Transitive confoliations

Let M be a manifold with a positive confoliation $\xi = \{\alpha = 0\}$. Let

$$H(\xi) = \{x \in M \mid \alpha \wedge d\alpha(x) > 0\}$$

be the contact part of M. Thus for a contact structure ξ we have $H(\xi) = M$.

For any subset $A \subset M$ the *saturation* $\hat{A}$ of A is the subset of M that can be reached from A by paths tangent to ξ, *i.e.* this is saturation in the sense of equivalence relations for the relation that two points can be connected by paths tangent to ξ.

Clearly if A is open then $\hat{A}$ is also open. If ξ is a foliation, $\widehat{\{x\}}$ is the leaf through x. A saturated set in a confoliation contains a neighborhood of any of its contact points — or in other words, the frontier of a saturated is contained in the fully foliated part of ξ. In particular, if ξ is contact and S is non-empty, then $\hat{S}$ is all of M (following the usual convention that manifolds are connected unless specifically stated otherwise.)

Just as for foliations, a *minimal set* of a confoliation ξ is defined to be a minimal non-empty closed saturated subset of ξ. Since the frontier of a closed saturated set is fully foliated, a minimal set of a confoliation is either the whole manifold, or a subset of the foliated part. A confoliation ξ on a compact 3-manifold M is called a *transitive* confoliation if it has no proper saturated subsets, or equivalently, if the saturation of its contact part is all of M. If $H(\xi)$ is not empty, this is also equivalent to the non-existence of a proper subset that is a minimal set. The class of transitive foliations was first described (without the name) by Steve Altschuler in [3].

PROPOSITION 2.7.1. *Any C^2-confoliation can be C^0-approximated by a transitive confoliation.*

PROOF. The strategy is to find ways to perturb ξ to be contact in neighborhoods, that intersect every minimal set of ξ. One way to think of this to eliminate minimal sets one by one, thereby monotonically decreasing the fully-foliated part of ξ until none is left and one has a transitive confoliation. In actuality, we will first analyze how to make perturbations in individual neighborhoods, then find a finite collection of disjoint neighborhoods whose union intersects every minimal set so we can make the perturbations simultaneously.

We begin with the case that ξ is a foliation, and first take up the case that ξ has non-trivial holonomy.

If M is a minimal set for ξ, then in light of remark 1.2.1 we can find a curve Γ whose holonomy is attracting on the positive side. Apply Proposition 2.5.1 to perturb ξ into a confoliation $\tilde{\xi}$ which is contact in a positive semi-neighborhood of Γ and diffeomorphic to ξ outside this neighborhood. Since every leaf of ξ is dense, $\tilde{\xi}$ is transitive.

Otherwise, the minimal sets of ξ consist of exceptional minimal sets and closed leaves.

Sacksteder's theorem (see Theorem 1.2.5 asserts that each exceptional minimal set has a curve with non-trivial linear holonomy. Choose one such curve for each exceptional minimal set, and perturb its neighborhood to contact using Proposition 2.6.1.

If there is any interval's worth or circle's worth of closed leaves, we can perturb ξ as a foliation to remove all but two closed leaves in any interval or circle, reducing to the case that every closed leaf has has non-trivial holonomy on both sides.

On any closed leaf S with non-trivial linear holonomy, there is some simple closed curve on S with non-trivial linear holonomy, and we can use Proposition 2.6.1 to perturb a neighborhood to be contact.

Otherwise, we will look a little harder at the holonomy around S. (We will also describe below an alternate technique that can be used more blindly, without analyzing the holonomy.) Pick a set of generators for $\pi_1(S)$. If we consider a point p near S on a normal arc where the holonomy action is not trivial, then we can view the holonomy under a high magnification that normalizes the holonomy action so that the generators displace p by a maximum distance of 1. Because of differentiability, under this magnification the holonomy around any generator is close to a translation. As analyzed in [**57**], this means holonomy commutes up to much smaller motion, and passing to the limit, we obtain a collection of asymptotic cohomology classes in $H^1(S, \mathbb{R}) = \mathrm{Hom}(\pi_1(S), \mathbb{R})$ that describe the limiting behavior of holonomy near the closed leaf S as scaled versions of groups of translations. One can picture this phenomenon by rescaling the $[-1, 1]$ coordinate in a neighborhood $S \times [-1, 1]$ of S using a highly repelling affine transformation whose fixed point p is close to S, using a scale factor such that everywhere near $S \times p$ the foliation is bounded from the vertical and somewhere near $S \times p$ bounded from horizontal. Any sequence of these rescalings has a subsequence that converges. In the limit, S recedes to infinity and we obtain a foliation on $S \times \mathbb{R}$ defined by a closed 1-form. The asymptotic cohomology classes are the restrictions of these limiting 1-forms to S. (This assumes that the initial coordinates $S \times [-1, 1]$ for a neighborhood were chosen as flat as possible, so that the leaves are as close to horizontal as the holonomy).

If these asymptotic cohomology classes do not lie on a single ray in $H^1(S)$, there is some homology class that pairs positively with an asymptotic class from above and pairs negatively with an asymptotic class from below. Choose any simple closed curve in such a homology class; it has sometimes attracting holonomy, and we can apply 2.5.1 c) to perturb a neighborhood to be contact.

If there is no simple curve on S with sometimes attracting holonomy, we will use another trick. Let $\gamma \in H^1(S)$ be a generator for the ray of asymptotic cocycles along S. Choose a flow ϕ_t on the unit interval with C^∞ contact to the identity at its endpoints, and construct a foliated I-bundle over S by using the action of $\pi_1(S)$ that takes an element $\alpha \in \pi_1(S)$ to $\phi_{-\gamma(\alpha)}$. Now cut open the foliation along S and put in a copy of the foliated I-bundle. We have replaced a single "indifferent" leaf by two leaves that have curves, γ_1 with weakly attracting holonomy and γ_2 with weakly repelling holonomy. This foliation may only be C^1-smooth: in particular, if the second derivative of holonomy is non-trivial on one side, that is incompatible with smooth weakly-attracting holonomy. This is not a real issue since it is soon to be a contact structure which is smoothable, but if we wish we can apply the flattening trick 1.2.11 to damp out all derivatives of holonomy on both sides, to obtain a foliation as smooth as the original. Now, apply 2.5.1 to perturb it to be a confoliation which is contact near γ_1 and γ_2.

Here is the alternative technique for finding a universal I-bundle for S, provided S is not a torus that when inserted in place of a leaf S with two-sided holonomy automatically creates sometimes-attracting holonomy; therefore it can be used without understanding the specific holonomy at all.[2] The idea is to modify the trivial foliation $S \times I$ by displacing leaves vertically along an infinite collection of disjoint transverse annuli that intersect every interior leaf and limit on both boundary components. This eliminates any interior closed leaves, and can be used to create sometimes attracting and sometimes expanding holonomy everywhere.

More explicitly: First, choose a doubly-infinite sequence $\Gamma_{,i}$ $i \in \mathbb{Z}$, of non-trivial homotopy classes of oriented simple closed closed curves on S, such that

- For all i the curves Γ_i and Γ_{i+1} are disjoint.
- Homotopy class of every oriented simple closed curve occurs as Γ_i infinitely often for both positive and negative $i \in \mathbb{Z}$, and moreover,
- Every pair of disjoint oriented simple closed curves occurs infinitely often as a pair (Γ_i, Γ_{i+1}) for both positive and negative $i \in \mathbb{Z}$.

Such a sequence exists in view of the fact, long known in surface topology, that one can get from any non-trivial simple closed curve to any other by a sequence of non-trivial simple closed curves such that adjacent curves in the sequence are disjoint.

Choose a cover of the interior of I by open intervals U_i, $i \in \mathbb{Z}$, such that for all i, U_i intersects U_{i+1}, but when $j - i > 1$ we have $U_j \cap U_i = \emptyset$.

Choose a family of C^∞ diffeomorphism ϕ_i supported in U_i such that for any $x \in U_i$, $\phi_i(x) > x$, but such that the ϕ_i tend C^∞ to zero very quickly as $|i| \to \infty$.

Now take the product $S \times I$ with the trivial foliation, slice open along the disjoint annuli $\Gamma_i \times U_i$, and glue the left side back to the right side with the diffeomorphism $id_{\Gamma_i} \times \phi_i$. The result is a C^∞ foliation provided the ϕ_i have been chosen to flatten out quickly enough. Since every leaf intersects an annulus, the resulting foliation has no closed leaves. For any non-separating simple closed curve α on S, there are pairs of disjoint curves (γ_1, γ_2) such that γ_1 intersects α once, but γ_2 does not intersect α. Therefore the copies of α on the two boundary components of $S \times I$ each have sometimes attracting and sometimes repelling holonomy on the interior side. When this foliation is inserted to replace a closed leaf, both replacement leaves have sometimes attracting holonomy, and we can proceed as before to insert a contact portion that intersects both of these closed leaves.

The union of minimal sets of ξ is compact, therefore there is a finite collection of neighborhoods as constructed whose union intersects all minimal sets and therefore all leaves of ξ and such that the foliation in each of these neighborhoods individually can be perturbed to be contact. However, it is likely that the individual perturbations conflict if the neighborhoods intersect.

One way to resolve the issue of disjointness would be to initially apply 1.2.10 to perturb ξ so that ξ has only finitely many closed leaves. However, we can deal with this issue more directly, since each type of neighborhood we have constructed can be shrunk as much as close as we like to its intersection with the minimal set under scrutiny. Explicitly, after an initial perturbation to remove any connected intervals or circles of closed leaves, we can cover the space of closed leaves by a disjoint union of foliated I-bundles $S \times I$, where the I-directions are short. An

---

[2]According to a theorem of Nancy Kopell [**38**], for any pair of commuting C^2 non-identity diffeomorphisms of I, a fixed point of one is a fixed point of the other. This implies that there is no "universal" C^2 foliated I-bundle for T^2. However, in the case of T^2, the previous construction that analyzes the cohomological direction of spiraling of holonomy is particularly elementary.

arbitrarily small neighborhood of any such I-bundle can be replaced by a foliated I-bundle as constructed above, by either technique, to make it transitive. If the I-bundles are thin, this is a C^0-small perturbation.

Consider now the case when ξ is a foliation without holonomy. According to Corollary 1.2.3 the foliation ξ can be approximated, in this case, by a fibration. Hence, using Proposition 1.2.10 we can further approximate ξ by a foliation with a finite number of closed leaves and no other minimal sets. Thus we return to the case of a foliation with holonomy already considered above. This finishes off the proof for the case of a foliation.

The case that ξ is a confoliation is very similar to that of foliations. As before, the idea is to modify ξ to be contact in small neighborhoods that intersect every minimal set.

Set $H = II(\xi)$, $\hat{H} = \widehat{H(\xi)}$, $C = M \setminus \hat{H}$ and suppose that H is not empty. First we observe that C coincides with the set $\mathrm{Fol}(\xi)$ which was defined in Section 1.2 above. Thus according to Proposition 1.2.12 the subset $C \subset M$ is closed, and (C, ξ) is a *lamination*.

One point which makes the cases of confoliations and foliations slightly different is that in the case of a confoliation the holonomy along a curve Γ depends on the choice of a transversal annulus.

Thus in this case the non-triviality of the holonomy along a curve Γ means that the holonomy is non-trivial for some choice of a transversal annulus. Notice, however, that if the restriction $\xi|_C$ has non-trivial *linear* holonomy along Γ (see 1.2) then this property holds for the holonomy along Γ of the full confoliation ξ *for any choice of a transversal annulus*. With these changes Proposition 2.5.1 and its proof hold for the case of a confoliation. The limits of rescaled holonomy near any closed leaf of a confoliation gives a collection of asymptotic cohomology classes on the leaf just like that for a foliation. Thus the proof of Proposition 2.7.1 can now be finished similarly to the case of foliations with holonomy which was considered above.

Of course, if the confoliation ξ is positive then the perturbation along the holonomy curve Γ can be done only into a positive confoliation. $\square$

2.8. Propagation of the perturbation along the leaves

The following Proposition 2.8.1 is due to Steve Altschuler (see [**3**]). Unlike ours, his proof was PDE-based. Namely, he used short time existence results for a linearized heat type equation associated to the problem. We use here a more direct geometric approach.

PROPOSITION 2.8.1. *Any C^k-smooth transitive confoliation, $k \geq 1$, can be C^k-approximated by a contact structure.*

To prove Proposition 2.8.1 we will need the following

LEMMA 2.8.2. *Let η be a C^k-confoliation in the domain*

$$V = \{|x|, |z| \leq 1, 0 \leq y \leq 1\}$$

given by a 1-form $dz - a(x, y, z)dx$. Suppose that the confoliation is contact near $\{y = 1\}$. Then η can be approximated by a confoliation $\hat{\eta}$ which coincides with η

together with all its derivatives along the boundary ∂V and which is contact inside V.

PROOF. We have

$$\frac{\partial a}{\partial y}(x, y, z) \geq 0 \quad \text{everywhere in } V$$

and

$$\frac{\partial a}{\partial y}(x, 1, z) > 0.$$

Thus there exists a function $\hat{a}(x, y, z)$ such that

$$\frac{\partial \hat{a}}{\partial y}(x, y, z) > 0 \quad \text{everywhere in int} V$$

and $\hat{a}$ coincides with a along ∂V with all its derivatives. Then the confoliation

$$\hat{\eta} = \{dz - \hat{a}(x, y, z)dy = 0\}$$

is the perturbation of η with the required properties. □

PROOF OF PROPOSITION 2.8.1. Let ξ be a positive transitive confoliation. Let $H = H(\xi)$ be the *hot zone* of ξ, i.e the subset of M where the confoliation ξ is a contact structure. For each point $p \in M$ we choose a curve Γ which begins at p and ends at a point $p' \in H$, and an embedding

$$F_p : V = I \times D_1 \times D_2 \to M,$$

where $D_1 = D_2 = [-1, 1]$, with the following properties

(i): $F_p(I \times 0 \times 0) = \Gamma$, $F_p(0 \times 0 \times 0) = p$, $F_p(1 \times 0 \times 0) = p'$;
(ii): $F_p(I \times z \times x)$ is tangent to ξ for all $z \in D_1$, $x \in D_2$;
(iii): $F_p(y \times D_1 \times x)$ is transversal to ξ for all $x \in D_2, y \in I$;
(iv): $F_p(1 \times D_1 \times D_2) \subset H$.

Denote by U and U' disks of radii $1/2$ and $3/4$ in $D_1 \times D_2$ centered at the origin. Set $W = \text{int}(I \times U)$, $W' = \text{int}(I \times U')$. We can choose finite number of points $p_1, \ldots, p_N \in M$ such that the open sets $W_i = F_{p_i}(W)$, $i = 1, \ldots, N$, cover the whole manifold M. Set $W_i' = F_{p_i}(W')$, $i = 1, \ldots, N$, $V_i = F_{p_i}(V)$. Thus we have $W_i \subset W_i' \subset V_i$. Let (y, z, x) be the coordinates in $I \times D_1 \times D_2$ which correspond to the splitting. Then the confoliation $\xi_i = (F_{p_i})^*(\xi)$ can be defined by a form

$$\alpha = dz - a_i(x, y, z)dx.$$

Thus we can apply Lemma 2.8.2 and C^k-perturb ξ_1 into a confoliation ξ_1' which is contact in W'. The push-forward of ξ_1' defines the required perturbation $\tilde{\xi}_1 = (F_{p_1})_*(\xi_1')$ in V_1.

Unfortunately, we cannot simply continue the process because the perturbation inside V_1 affects the properties of the embeddings F_{p_i} for $i = 2, \ldots, N$. However, if the perturbation is sufficiently small, then it is possible to modify slightly the embeddings F_{p_i}, $i = 2, \ldots, N$ into F_{p_i}', to satisfy the conditions (ii)-(iv), and to have the following condition

(i'): $F_{p_i}'(W') \supset W_i$, $i = 2, \ldots, N$,

instead of (i). But this is sufficient to continue inductively the process of perturbation. □

PROOF OF THEOREM 2.4.1. Theorem 2.4.1 is a direct corollary of propositions 2.7.1 and 2.8.1.   □

More attentive analysis of the proof of Proposition 2.8.1 allows to prove the following more quantitative and parametric version 2.8.3 of Proposition 2.8.1 which is needed for the proof of Theorem 2.1.2.

Let us fix a Riemannian metric on M. Then with a co-oriented plane field ξ we can canonically associate a 1-form α such that $\xi = \{\alpha = 0\}$, and the point-wise norm of α equals 1. Define a function $N_\xi : M \to \mathbb{R}_+$ by the formula $N_\xi = ||\alpha \wedge d\alpha||$.

Let ξ be a confoliation and $K \subset H(\xi)$ be a compact set in the contact part of ξ, and let $\hat{K}$ be the saturation of K. For any point $q \in M$ we denote by $\mathrm{dist}_\xi(q, K)$ the Carnot-Caratheodory distance from q to K, i.e.

$$\mathrm{dist}_\xi(q, K) = \inf_\gamma (\mathrm{length}(\gamma)),$$

where γ is a curve tangent to ξ which connects q with a point of K. We set $\mathrm{dist}_\xi(q, K) = \infty$ if q cannot be reached from K along a path tangent to ξ.

Set

$$\mathrm{dist}_\xi(K) = \sup_{q \in M} \mathrm{dist}_\xi(q, K)).$$

Thus $\mathrm{dist}_\xi(K) < \infty$ if and only if $\hat{K} = M$.

PROPOSITION 2.8.3. *Let $\xi_t, t \in [0,1]$, be a family of confoliations on a compact manifold M beginning with the foliation ξ_0. Let K_1 and K_2 be compact subsets of M such that $K_1 \subset \mathrm{int}\, K_2 \subset K_2 \subset H(\xi_t)$ for all $t > 0$. Suppose that $\mathrm{dist}_{\xi_t}(K_1) < \infty$ for all $t \in [0,1]$. Then there exists a smooth family of confoliations $\tilde{\xi}_t$ such that*

- $\tilde{\xi}_0 = \xi_0$;
- $\tilde{\xi}_t$ *is a contact structure for all $t > 0$;*
- $\tilde{\xi}_t$ *coincides with ξ_t on K_1 for all $t \in [0,1]$;*
- *we have the inequality*

(2.17) $$N_{\tilde{\xi}_t} \geq \frac{C \min_{K_2} N_{\xi_t}}{(1 + \mathrm{dist}_{\xi_t}(K_2))},$$

where the constant C depends only on the metric and the compact sets K_1, K_2 (and not on the family ξ_t).

PROOF OF THEOREM 2.1.2. If our C^2-foliation is minimal and has non-trivial holonomy then according to Ghys' theorem 1.2.7 it has a curve with non-trivial linear holonomy. If it is not minimal, then the closure of any leaf contains either a minimal exceptional leaf or a closed leaf. Any C^2-foliation with holonomy is either minimal, or has finite number of exceptional minimal sets and each of them has a curve with non-trivial linear holonomy (see 1.2.4 and 1.2.5 above). Moreover, the condition on closed leaves guarantees that there are only finitely many closed leaves. Thus, in all cases there exists a finite number of curves $\Gamma_1, \ldots, \Gamma_k$ with non-trivial linear holonomy such that any leaf of the foliation $\xi = \{\alpha = 0\}$ contains one of these curves in its closure. According to Proposition 2.6.1 there exists a neighborhood $U \supset \Gamma = \bigcup_1^k \Gamma_i$ such that $\xi|_U$ can be linearly deformed into a contact structure. This means that there exists a 1-form β such that $\langle \alpha, \beta \rangle|_U > 0$ and such that $\beta|_{M \setminus U} \equiv 0$. Choose smaller neighborhoods $U', U'' \supset \Gamma$ such that $\overline{U}' \subset U'' \subset \overline{U}'' \subset U$. Apply now Proposition 2.8.3 to the family of confoliations $\xi_t = \{\alpha + t\beta = 0\}$ on M, and to $K_1 = \overline{U}'$, $K_2 = \overline{U}''$. Let $\tilde{\xi}_t = \{\tilde{\alpha}_t = 0\}$ be the family of contact structures provided by this proposition. Then $\tilde{\xi}_0$ coincides with the foliation ξ, and the the norm $N_{\tilde{\xi}_t}$ satisfies the inequality (2.17). On the other hand the distances $\mathrm{dist}_{\xi_t}(K_2), t \in [0,1]$, are uniformly bounded above by a constant C_1, and we have

$$N_{\xi_t}|_{K_2} \geq C_2 t \min_{K_2} \langle \alpha, \beta \rangle \geq C_3 t,$$

for constants $C_2, C_3 > 0$. Thus,

$$N_{\tilde{\xi}_t} \geq C_4 t, \ C_4 > 0,$$

and therefore

$$\frac{d(\tilde{\alpha}_t \wedge d\tilde{\alpha}_t)}{dt} > 0,$$

which by definition means that $\tilde{\xi}_t, t \in \mathbb{R}$, is a linear deformation of the foliation ξ into a contact structure.   □

2.9. Discussion

2.9.1. Non-coorientable case. Note that Theorem 2.4.1 remains true in the non-coorientable case as well. The proof follows the same scheme with the additional observations and adjustments which we describe below.

Let (M, ξ) be a non-co-orientable foliation on the orientable manifold M. There exists a double cover $\pi : \tilde{M} \to M$ so that the induced foliation $\tilde{\xi}$ on $\tilde{M}$ is co-orientable.

If $(\tilde{M}, \tilde{\xi})$ is a foliation without holonomy which admits a co-orientation reversing involution then it must have a closed leaf. Indeed, the universal covering $(\widehat{M}, \hat{\xi})$ of the foliation $\tilde{M}, \tilde{\xi}$ is a foliation on $\mathbb{R}^2 \times \mathbb{R}$ transversal to the fibers $q \times \mathbb{R}, q \in \mathbb{R}^2$. The co-orientation reversing involution $i : \tilde{M} \to \tilde{M}$ lifts to a co-orientation reversing involution $I : \widehat{M} \to \widehat{M}$, which commutes with the deck transformations $F_i : \widehat{M} \to \widehat{M}, i = 1, \ldots, k$. The involution I induces a reflection $\bar{I} : \mathbb{R} \to \mathbb{R}$ whose fixed point gives rise to a leaf of $\hat{\xi}$ invariant under all deck transformations $F_i : \widehat{M} \to \widehat{M}, i = 1, \ldots, k$. Thus its the quotient is a closed leaf of the foliation $\tilde{\xi}$. But Theorem 1.2.2 implies that the foliation $(\tilde{M}, \tilde{\xi})$ is a fibration, and thus the quotient (M, ξ) has plenty of closed co-orientable leaves. Therefore in this case one can proceed as in the proof of 2.4.1 to create a curve with non-trivial holonomy.

If a leaf $\tilde{S}$ of $(\tilde{M}, \tilde{\xi})$ has a curve with non-trivial holonomy then the leaf F of (M, ξ), covered by $\tilde{F}$, has a curve Γ whose covering curve $\tilde{\Gamma}$ has non-trivial holonomy. Suppose that the co-orientation is being reversed along the curve Γ. Then the splitting

$$U = \Gamma \times D_1 \times D_2$$

of a neighborhood of Γ can be chosen in such a way that the projection $\pi|_U$ identifies the orbits of the involution

$$(x, y, z) \overset{i}{\mapsto} (\tilde{x}, -y, -z),$$

where $x \mapsto \tilde{x}$ is a fixed point free involution on Γ. Our foliation ξ is invariant under the involution i. Thus our goal is to make the perturbations constructed in 2.5.1 commuting with i. Notice that the case 2.5.1b) cannot occur because of the symmetry of the foliation ξ under the involution i. Hence we can consider here only the cases a) and c) of 2.5.1. Let $F : U \cap \{-1 \leq y \leq -1/2\} \to U \cap \{-1 \leq y \leq -1/2\}$

be the diffeomorphism constructed in the proof 2.5.1a) and c). We extend F to $U \cap \{1/2 \leq y \leq 1\}$ by the formula $F(u) = i(F(i(u)))$, $u \in U \cap \{1/2 \leq y \leq 1\}$. Then as in the proof of 2.5.1 the foliation $F_*(\xi)$ can be extended to U as an i-invariant contact structure on $\Gamma \times (-1/2, 1/2) \times (-1, 1)$. The passage from a transitive confoliation to a contact structure works in the non-co-orientable case without any changes.

2.9.2. Deformation vs. perturbation, and the problem of smoothness. A slight modification of the argument in Theorem 2.4.1 allows to prove the following result.

THEOREM 2.9.1. *Given a C^0-foliation ξ one can find a family of smooth confoliations ξ_t, $t \in [0, 1]$ which continuously depends on the parameter t, such that $\xi_0 = \xi$, ξ_1 is a contact structure, and ξ_t is C^0-close to ξ for all $t \in [0, 1]$.*

Analyzing the proof of Theorem 2.4.1 we see that there were two instances where we could not get a better than C^0-approximation. The first one was when we needed to approximate the foliation without holonomy by a foliation with holonomy, and the second one when we created a contact zone near a curve on a closed leaf which has one-sided attracting holonomy, or sometimes attracting two-sided holonomy. On the other hand, according to 2.6.2 if the holonomy along a curve is nice then one can create a contact zone near this curve via a smooth *deformation*. Thus, we have

PROPOSITION 2.9.2. *Suppose a C^k-confoliation ($k \geq 2$) ξ has holonomy, or $H(\xi) \neq \emptyset$. Suppose also that if a closed leaf S of ξ has non-trivial holonomy on one of its sides then there exists a curve $\Gamma \subset S$ with nice holonomy on the same side. Then ξ can be C^k-deformed into a contact structure.*

COROLLARY 2.9.3. *Suppose that a C^k-confoliation (M, ξ), $k \geq 2$, admits an exact 2-form ω such that $\omega|_\xi \neq 0$ (comp. Section 3.2 below.) Then ξ can be C^k-deformed to a contact structure.*

PROOF. Existence of the exact form ω prohibits closed leaves of ξ because for each such leaf S we would have $\int_S \omega \neq 0$. Similarly, ξ cannot be a foliation without holonomy because otherwise it could be C^0-perturbed to a foliation with closed leaves. Hence we can apply 2.9.2. □

We believe that in the general case the approximation can be, probably, improved at least to the class C^1.

Observe also that we twice used the fact that the original foliation is C^2-smooth. The first time it was again for perturbations of foliations without holonomy, and the second time to apply Sacksteder's theorem. In particular, we have

PROPOSITION 2.9.4. *If a C^1-confoliation (or even C^0-foliation) has holonomy, or if $H(\xi) \neq \emptyset$, and if ξ is C^2-smooth in the complement of the union $C(\xi)$ of closed leaves, then it can be C^0-approximated by a contact structure.*

However, it seems feasible that the result holds without any assumptions about the smoothness of the foliation ξ.

CHAPTER 3

Taut vs. Tight

3.1. Tight contact structures and taut foliations

A contact structure ξ on a 3-manifold M is called *overtwisted* (see [11]) if there exists an embedded disk $D \subset M$ such that ∂D is tangent to ξ but D itself is transversal to ξ along ∂D (see Figure 3.1).

A contact structure is called *tight* if it is not overtwisted.

As it was shown in [11], [12], [24] tight and overtwisted structures constitute two completely different worlds. Overtwisted structures are very flexible and abide an h-principle: the isotopy classification of overtwisted structures coincides with their homotopy classification as tangent plane distributions (see [11]). On the other hand tight contact structures exhibit a lot of rigidity properties. For instance, on some manifolds (like $S^3, \mathbb{R}P^3, \mathbb{R}^3, S^2 \times S^1$) there exists a unique, up to isotopy, contact structure (see [12]). Another manifestation of rigidity of tight contact structures is the inequality which we formulate in Theorem 3.3.1 below.

Moving to the other end of the confoliation scale, *i.e.* to the foliations, let us recall that a foliation ξ, given on a closed manifold M, is called *taut* (see [20]) if it is different from the foliation ζ on $S^2 \times S^1$, and satisfies any of the following equivalent (for a closed manifold M) properties (their equivalence is a corollary of the results of S.P. Novikov and D. Sullivan, see [48] and [52]):

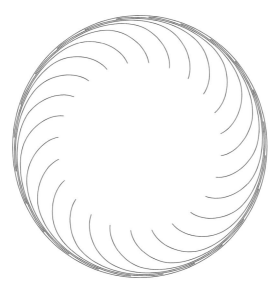

FIGURE 3.1. Characteristic foliation near the boundary of an overtwisted disk.

O_1: each leaf is intersected by a transversal closed curve;

O_2: there exists a vector field X on M which is transversal to ξ and preserves a volume form Ω on M;

O_3: M admits a Riemannian metric for which all leaves are minimal surfaces.

A taut foliation cannot have Reeb components. In fact, a necessary and sufficient condition for tautness is that the foliation has no *generalized Reeb components* (see [28]). A generalized Reeb component is a submanifold $N \subset M$ of maximal dimension bounded by tori, such that the orientation of these tori as leaves of the foliation is the same as (or simultaneously opposite to) their orientation as the boundary components of N.

3.2. Symplectic filling

Let (M, ξ) be a confoliation and ω a closed 2-form on M. We say that ω *dominates* ξ if $\omega|_\xi$ does not vanish.

For a foliation the existence of a dominating form ω is equivalent to the definition O_2 of a taut foliation. Indeed, if the vector field X is transverse to ξ and preserves Ω then the closed 2-form $\omega = X \rfloor \Omega$ dominates ξ. Conversely, suppose ω dominates $\xi = \{\alpha = 0\}$. Then the vector field X, such that $X \rfloor \omega = 0$ and $\alpha(X) = 1$, is transverse to ξ and preserves the volume form $\Omega = \omega \wedge \alpha$.

Thus, the definition O_2 of tautness can be rewritten in the following form:

O_2': ξ admits a closed dominating 2-form.

Suppose now that a 3-manifold M with a positive confoliation ξ bounds a compact symplectic 4-manifold (W, ω). We call (W, ω) a *symplectic filling* of the confoliated manifold (M, ξ) if $\omega|_M$ dominates ξ and M is oriented as the boundary of the canonically oriented symplectic manifold (W, ω). Of course, when ξ is a foliation then the orientation condition is irrelevant.

A confoliated manifold which admits a symplectic filling is called *symplectically fillable*. It is called *symplectically semi-fillable* if it is a connected component of a symplectically fillable confoliated manifold.

Notice that a compact complex manifold W of (complex) dimension 2 with strictly pseudo-convex boundary is automatically Kähler, and therefore symplectic. Moreover, the symplectic form dominates the contact structure ξ formed by the complex tangent lines to the boundary $M = \partial W$ (see Section 1.1.5 above). The orientation condition is also met, which means that the contact structure ξ is symplectically fillable. Thus CR-structures on strictly pseudo-convex boundaries of compact complex manifolds provide a rich source of examples of symplectically fillable contact structures.

REMARK 3.2.1. Pseudo-convex boundaries of compact complex manifolds have to be connected (as it follows, for instance, from Lefshetz's theorem). However, a compact symplectic manifold may have a disconnected contact boundary. This fact motivates the notion of symplectic semi-fillability. The first example of disconnected boundary was constructed by D. McDuff in [43]. Her example was generalized in [22] and [46]. A large class of examples is provided by Corollary 3.2.5 below.

PROPOSITION 3.2.2. *Taut foliations are symplectically semi-fillable.*

PROOF. Let ω be a dominating 2-form for the taut foliation $\xi = \{\alpha = 0\}$ on M. Set $W = M \times [0,1]$, and define a closed 2-form $\tilde{\omega} = p^*\omega + \varepsilon d(t\alpha)$, where p is the projection $W \to M$. When $\varepsilon > 0$ is small then the form $\tilde{\omega}$ is non-degenerate and dominates ξ on $\partial W = M \times 0 \cup M \times 1$. $\qquad\square$

Notice here that the orientation defined by $\tilde{\omega}$ on the boundary of W coincides with the orientation of M on $M \times 1$, and is opposite to it on $M \times 0$.

REMARK 3.2.3. It is possible that all taut foliations are symplectically fillable, and not only semi-fillable. However, this is unknown.

As was proven in [30] and [13]

THEOREM 3.2.4. *Symplectically semi-fillable contact structures are tight.*

COROLLARY 3.2.5. *Contact structures, C^0-close to a taut foliation, are symplectically semi-fillable, and, therefore, tight.*

PROOF. Symplectic fillableness is a C^0-open condition. $\qquad\square$

It seems useful to extend the notion of symplectic (semi-)fillabillity to confoliations on non-compact manifolds.

Following M. Gromov ([30]), we say that a symplectic manifold (W,ω) has *finite geometry at infinity* if there there exists an almost complex structure J on W such that:

- J is calibrated by ω, i.e. $\omega(v, Jv) > 0$ and $\omega(Jv, Jv) = \omega(v, v)$ for any non-zero tangent vector $v \in T(W)$;
- the corresponding Riemannian metric $\omega(\cdot, J\cdot)$ is complete, has bounded above sectional curvature, and bounded away from zero injectivity radius.

For instance the universal cover of a closed symplectic manifold has finite geometry at infinity. Another class of examples is provided by Stein complex manifolds endowed with their canonical symplectic structure (see [17]).

We say that a confoliation ξ on a possibly non-compact 3-manifold M is semi-fillable if there exists a symplectic 4-manifold (W,ω) of finite geometry at infinity, such that

1. M is the boundary of W;
2. ξ is J-invariant for an almost complex structure J implied by the definition of finite geometry at infinity;
3. M is a (possibly weakly) J-convex boundary of W.

The latter condition is equivalent to the orientation condition which we formulated in the compact case.

REMARK 3.2.6. If M is one of the boundary components of a symplectic manifold (W,ω) which satisfies the above properties 1-3 then one can strip off other boundary components of W and by properly modifying ω and J near these components, ensure that the new symplectic manifold (W',ω') satisfy the same properties. Hence our definition of filling in the non-compact case is an analog of semi-fillabillity, and not fillabillity, in the compact case.

Theorem 3.2.4 remains true in the non-compact case with this definition of symplectic semi-fillabillity. Similarly, Proposition 3.2.2 can be modified as follows

PROPOSITION 3.2.7. *A universal cover of a taut foliation is symplectically semi-fillable.*

In particular, we can strengthen Corollary 3.2.5 (see also the discussion in Section 3.5.2 below):

COROLLARY 3.2.8. *Contact structures, C^0-close to a taut foliation, have tight universal covers.*

In Section 3.5.4 (see Theorem 3.5.12 below) we prove a stronger statement asserting tightness of contact structures, and even their universal covers, which are C^0-close to foliations without Reeb components.

The following theorem of D. Gabai (see [20]) shows that there are a lot of closed 3-manifolds which admit taut foliations. Namely, he proved

THEOREM 3.2.9. (D. Gabai, [20]) *Let M be an irreducible 3-manifold with non-trivial homology group $H_2(M)$. Let $F \subset M$ be an oriented surface which represents a non-trivial homology class and has minimal genus among surfaces representing the same class. Then there exists a taut foliation ξ on M which has F as one of its leaves.*

REMARK 3.2.10. In [14] Eliashberg described a class of index 1 and 2 surgeries of contact manifolds (see also Gompf's paper [27], where the surgery technique is further developed). These surgeries have the following important property: if the original contact manifold is symplectically (semi-)fillable then after the surgery it is symplectically (semi-)fillable as well. However, if the original manifold is only tight it is unknown whether it remains tight after surgeries of index 2.[1] Proposition 3.2.2 shows that performing contact surgeries on contact manifolds which are deformations of taut foliations, we remain in the class of symplectically fillable and, therefore, tight contact structures.

From 3.2.9, 2.4.1, 3.2.5 and 3.2.10 we conclude

COROLLARY 3.2.11. *Any 3-manifold M with $H_2(M)/h(\pi_2(M)) \neq 0$, where $h : \pi_2(M) \to H_2(M)$ is the Hurewicz homomorphism, admits a tight (moreover symplectically semi-fillable) contact structure.*

3.3. The inequality

It sounds surprising that while the definitions of a tight contact structure and a taut foliation look completely unrelated, they both satisfy the same remarkable inequality which was independently proven in the case of taut foliations by Thurston in [53] (see also [56]) and in the case of tight contact structures by Bennequin and Eliashberg (see [7] and [12]).

Let ξ be an orientable tangent plane distribution on an oriented 3-manifold M, and $F \subset M$ an orientable 2-surface which is either closed or has a boundary Γ transversal to ξ. If F is closed then we orient F and ξ arbitrarily, but if $\Gamma \neq \emptyset$ then the orientations of M, F and ξ should be related in the following way:

---

[1]Tightness is preserved by index 1 surgeries. This was established independently by V. Colin ([9]) and S. Makar-Limanov ([42]).

When Γ is oriented as the boundary of the oriented surface F, then at a point $x \in \partial F$ the orientation of the plane $\xi(x)$ together with the orientation of $T_x(\Gamma)$ should coincide with the orientation of the tangent space $T_x(M)$.

We denote by $\chi(F)$ the Euler characteristic of F and by $e(\xi)[F]$ the value of the Euler class $e(\xi) \in H^2(M)$ evaluated on F in the case when F is closed, and the relative Euler number of $\xi|_F$ if $\Gamma \neq \emptyset$. The relative Euler number can be defined as follows. Consider a vector field X along Γ which generates the line field $\xi \cap T(F)$. Then $e(\xi)[F]$ is the obstruction for the extension of X to F as a vector field in ξ.

Notice that the number $e(\xi)[F]$ remains unchanged when the orientations of ξ and F are simultaneously reversed. In particular, in the case of a surface with boundary our orientation agreement shows that $e(\xi)[F]$ is determined by the choice of the orientation of M.

THEOREM 3.3.1. *Let ξ be a tight positive contact structure or a taut foliation on an oriented 3-manifold M. If F is a closed embedded orientable 2-surface $F \subset M$ which is different from S^2 then the following inequality holds:*

(3.1) $$|e(\xi)[F]| \leq -\chi(F).$$

If $F = S^2$ then

(3.2) $$e(\xi)[F] = 0.$$

If F is a surface with boundary transversal to ξ then we have the inequality

(3.3) $$e(\xi)[F] \leq -\chi(F).$$

The inequality (3.1) implies (see [13])

COROLLARY 3.3.2. *For any closed manifold M only finitely many cohomology classes from $H^2(M)$ can be represented as Euler classes of tight contact structures.*

Indeed, any integer homology class from $H_2(M)$ can be represented by a closed oriented surface. Thus the inequality (3.1) gives bounds for the possible values of the cohomology class $e(\xi)$ on the generators of $H_2(M)$.

It is still unknown whether only a finite number of homotopy classes of plane fields can be represented by tight contact structures. However, recently Kronheimer and Mrowka [39] proved, using the Seiberg-Witten theory that *symplectically semi-fillable contact structures may represent only a finite number of homotopy classes of plane fields*. This result, together with 2.4.1 and 3.2.5 implies

COROLLARY 3.3.3. *Only finitely many homotopy classes of tangent plane fields are representable by taut foliations.*

E. Giroux ([24] and Y. Kanda ([37]) proved that there are infinitely many non-diffeomorphic tight contact structures on the 3-torus T^3, with all of them representing the same homotopy class of plane fields. These examples can be thought of as starting with the standard contact structure for T^3 as the unit cotangent bundle of T^2, then pulling back under covering maps $T^3 \to T^3$ unwiding the fiber. It is likely, although unproven, that any 3-manifold which contains an irreducible 2-torus, admit infinitely many non-isomorphic tight contact structures.

QUESTION 3.3.4. *Which 3-manifolds admit infinitely many non-diffeomorphic tight contact structures?*

It is interesting to point out that the torus T^3 has also infinitely many *diffeo-morphic but non-isotopic* contact structures (see [**18**] and [**24**]).

The natural question in connection with theorem 3.3.1 is to describe the general class of *confoliations* for which the above inequalities hold. Let us denote by $\mathcal{T}$ the class of all confoliations which satisfy the relations (3.1)–(3.3). The intersection of the class $\mathcal{T}$ with the class of contact structures consists exactly of tight contact structures.

On the other hand, in the case of foliations the class $\mathcal{T}$ contains more than just taut foliations. In fact, Thurston proved in [**53**] the inequalities (3.1)–(3.3) for a larger class of foliations, namely for foliations without Reeb components.

We can find further examples by applying the following elementary and obvious fact:

PROPOSITION 3.3.5. *The class $\mathcal{T}$ is closed in the C^0-topology.*

We can apply this proposition to the process of "turbularization". Given a foliation ξ, one can choose a closed transverse curve α, and transport the foliation by an isotopy that looks like a laminar flow supported in a regular neighborhood of α. It is possible for the flow to travel around α infinitely often while changing the foliation only a bounded amount, so that in the limit, one obtains a foliation with a Reeb component inserted near α (see Figure 3.4, also the discussion in the proof of Proposition 3.6.2). According to Proposition 3.3.5, if the original foliation was in $\mathcal{T}$, so is the new one.

The limit foliation has *vanishing cycles*, that is, non-trivial simple closed curves tangent to ξ that bound disks in the manifold which are related in spirit to violations of (3.3). Perhaps the definition of class $\mathcal{T}$ is biased in favor of contact structures, since in the case of a contact structure any curve tangent to ξ is approximable by curves transverse to ξ.

We will shortly define (in 3.5.1) another property that bans vanishing cycles. Nonetheless, the power of this condition for contact structures suggests that class $\mathcal{T}$ may have significance for foliations and confoliations as well.

Here is another example:

EXAMPLE 3.3.6. Let $M = S^3$, and ρ be the Reeb foliation. Actually, up to an orientation preserving homeomorphism, there are two different Reeb foliations (notice that there is a continuum of *non-diffeomorphic* Reeb foliations). To distinguish between them let us choose a co-orientation of ρ and pick positively oriented closed transversals T_1 and T_2 in each of the Reeb components of the foliation. We call the Reeb foliation positive or negative depending on the sign of the linking number $l(T_1, T_2)$. These two Reeb foliations will be denoted by ρ_+ and ρ_-. We leave it as an exercise to the reader to check that

the positive Reeb foliation ρ_+ belongs to the class $\mathcal{T}$ while the negative one does not.

It is interesting to observe that a perturbation of ρ_+ into a positive contact structure (which exists according to Theorem 2.4.1) is tight while its perturbation into a negative contact structure is overtwisted.

3.4. Contact geometry of planes in the standard contact $\mathbb{R}^3$

3.4.1. Simple planes in $\mathbb{R}^3$. As it was already mentioned above a germ of a contact structure along a surface is determined by the difeomorphism class of the characteristic foliation induced by on this surface by the contact structure.

A properly embedded plane in a contact manifold is called *simple* if it is transversal to the contact structure and carry the *standard* characteristic foliation, i. e. the foliation diffeomorphic to the fibration of $\mathbb{R}^2$ by parallel straight lines. For instance, the plane

$$\{y = 0\} \subset (\mathbb{R}^3, \{dz = ydx\})$$

is simple. We call it the *standard simple plane*.

QUESTION 3.4.1. *When a simple plane in the standard contact $(\mathbb{R}^3, \{dz = ydx\})$ is contactomorphic to the standard simple plane via a global contactomorphism of $\mathbb{R}^3$, or via a contactomorphism defined on a closed half-space bounded by this plane?*

We give below (see 3.4.3) a criterion for a simple plane to be standard. On the other hand, we also show (see 3.4.2) that non-standard simple planes do exist. We will also consider contact structures on the closed half-space $\mathbb{R}^3_+$ bounded by simple planes.

Given a function $\varphi : \mathbb{R}^2 \to [0, \infty)$, we denote by S_φ its graph $\{y = \varphi(x, z)\} \subset \mathbb{R}^3$. If the function φ has a compact support then the holonomy along the leaves of the characteristic foliation on S_φ defines a compactly supported diffeomorphism $h_\varphi : \mathbb{R} \to \mathbb{R}$. Namely,

$$h_\varphi(z) = \lim_{x \to +\infty} z(x),$$

where $z(x)$ is the solution of the differential equation $\frac{dz}{dx} = \varphi(x, z)$, which satisfies the initial condition $\lim_{x \to -\infty} z(x) = z$. If the support of φ is not compact but the function φ decays sufficiently fast when $|x| \to \infty$ (say, $\varphi(x, z) < \frac{C}{x^2}$), then the holonomy diffeomorphism h_φ is still defined, and although its support may not be compact, it has bounded displacement:

$$0 \leq h_\varphi(z) - z < 2C.$$

In the rest of this section we always consider the functions φ which satisfy this decay condition, so that the holonomy map h_φ is defined and has a bounded displacement. We set

$$\delta_\varphi = \sup_{z \in \mathbb{R}}(h_\varphi(z) - z).$$

3.4.2. Non-standard simple planes in $\mathbb{R}^3$. Suppose that a function φ is positive, and consider the domain $\Omega_\varphi = \{0 \leq y < \varphi(x, z)\}$.

PROPOSITION 3.4.2. *If $\delta_\varphi < +\infty$ then the domain Ω_φ is not contactomorphic to the half-space $\mathbb{R}^3_+ = \{y \geq 0\}$. Moreover, the standard contact $\mathbb{R}^3_+$ cannot be embedded into Ω_φ in such a way that the plane $\{y = 0\}$ is being mapped onto itself.*

PROOF. Suppose that there exists a contact embedding $f : \mathbb{R}^3_+ \to \Omega_\varphi$ which maps the plane $\{y = 0\}$ onto itself. We can assume that $f|_{\{y=0\}} = \mathrm{Id}$ because any diffeomorphism $\mathbb{R}^2 \to \mathbb{R}^2$ which preserves the standard (characteristic) foliation extends to a contact automorphism of $\mathbb{R}^3_+$. Take a compactly supported function $\alpha : \mathbb{R}^2 \to \mathbb{R}_+$ and consider the corresponding holonomy diffeomorphism $h_\alpha : \mathbb{R} \to \mathbb{R}$. Clearly, any compactly supported diffeomorhism $\mathbb{R} \to \mathbb{R}$ can be realized as the holonomy diffeomorphism h_α for an appropriate choice of the function α. Suppose that α is chosen in such a way that

$$\delta_\alpha = \sup_{z \in \mathbb{R}} (h_\alpha(z) - z) > \delta_\varphi .$$

Consider the image $T_\alpha = f(S_\alpha) \subset \Omega_\varphi$. Let $B \subset \mathbb{R}^2$ be a sufficiently large ball, and $\tilde{\varphi} : \mathbb{R}^2 \to \mathbb{R}$ a positive function such that:

- $\tilde{\varphi} < \varphi$;
- S_α is flat outside of the cylinder $C = B \times \mathbb{R} = \{(x, y, z) \mid (x, z) \in B\}$;
- the image $f(S_\alpha \cap C)$ is contained in $C \cap \{0 < y < \tilde{\varphi}(x, z)\}$.

Let U be the closed domain bounded by T_α and $S_{\tilde{\varphi}}$. Set $\tilde{U} = U \cap B$. It is clear from the construction, that the characteristic foliation on the sphere ∂U contains, after smoothing the corners, a non-singular closed leaf. But this contradicts to the tightness of the contact structure on Ω_φ.   □

Let us recall that all tight contact structures on $\mathbb{R}^3$ are isomorphic (see [16]). In particular, the open domain $\{y < \varphi(x, z)\}$ with the induced contact structure is contactomorphic to the standard contact $\mathbb{R}^3$. The image of the plane $\{y = 0\}$ under this contactomorphism divides $\mathbb{R}^3$ into two closed subspaces, one of which is standard but the other is not. Similarly, the open domain $\{-\varphi(x, z) < y < \varphi(x, z)\}$ is also contactomorphic to the standard contact $\mathbb{R}^3$. Therefore, we get a simple plane in $\mathbb{R}^3$ which divides $\mathbb{R}^3$ into two half-spaces, such that neither of them is isomorphic to the standard contact $\mathbb{R}^3_+$.

3.4.3. Convex planes: a criterion for a simple plane to be standard.

Any simple plane admits a transversal contact vector field. For instance, for the standard simple plane one can take $v = y\frac{\partial}{\partial y} + z\frac{\partial}{\partial z}$. However, sometimes this field cannot be extended to a complete (i.e. both directions integrable) contact vector field on the whole $\mathbb{R}^3$, or at least to the half-space bounded by this plane.

A simple plane P in a contact manifold is called *convex* (see [17] and [23] if it admits a complete globally defined contact vector field transversal to P. A co-oriented simple plane is called *semi-convex* if it admits a forward integrable contact vector field which is transversal to the boundary, and pointed at the direction defined by the co-orientation.

PROPOSITION 3.4.3. *Any contact half-space bounded by a simple semi-convex plane is isomorphic to the standard contact half-space.*

Before proving Proposition 3.4.3 let us recall some known results from the 3-dimensional contact topology.

SUMMARY 3.4.4.   **A:** If the characteristic foliations on two surfaces S_1 and S_2 are homeomorphic, then by a C^1-small isotopy of one of them, the characteristic foliations can be made diffeomorphic, so that the germs of contact structures on these surfaces become isomorphic. (See [**11**]).

B: Given a sphere S in a tight contact manifold, one can deform it by a C^0-small isotopy to a sphere $\tilde{S}$ whose characteristic foliation has exactly two elliptic singular points of opposite signs and no other singularities or limit cycles. In other words, the characteristic foliation on $\tilde{S}$ is homeomorphic to the characteristic foliation of a round sphere in the standard contact $\mathbb{R}^3$. Similarly, let D be a disk in a tight contact manifold whose boundary $\Gamma = \partial D$ is transversal to the contact structure. Suppose that $e(\xi)[D] = -1$ (see Section 3.3 above for the definition). Then D can be C^0-deformed via a contact isotopy fixed near the boundary in order to kill all singularities of the characteristic foliation, except one elliptic point. This is done in [**12**] using the elliptic-hyperbolic cancellation lemma from [**23**]. Given a torus T in a tight contact manifold one can perturb T by a C^0-small isotopy to obtain a torus T' transversal to the contact structure ξ and such that the characteristic foliation induced by ξ on the torus T' admits a circle S which transversally intersects all leaves of the foliation (see [**41**], [**24**]).

C: Any two tight contact structure on the spherical annulus $S^2 \times I$, which induce the same characteristic foliations on the boundary, are isotopic via an isotopy fixed at the boundary $S^2 \times 0 \cup S^2 \times 1$ (see [**12**]).

As a straightforward application of the statement C we get the following

LEMMA 3.4.5. *Let ξ and $\tilde{\xi}$ be two tight contact structures on the half-space $\mathbb{R}^3_+$, which coincide along the boundary $P = \partial \mathbb{R}^3_+$. Let $S_i \subset \mathbb{R}^3_+$, $i = 1, \ldots,$ be a sequence of embedded planes, and $D_i \subset P$, a sequence of discs such that*

- $S_i \cap P = P \setminus \operatorname{Int} D_i$;
- $\bigcup_i D_i = P$, $\bigcup_i B_i = \mathbb{R}^3_+$; $D_i \subset \operatorname{Int} D_{i+1}$, $B_i \subset \operatorname{Int} B_{i+1}$, $i = 1, \ldots,$. *Here we denote by B_i the ball bounded by the discs D_i and $\Delta_i = S_i \setminus (P \setminus D_i)$:*
- *For all $i = 1, \ldots$ the characteristic foliations induced by the contact structures ξ and $\tilde{\xi}$ on discs Δ_i are diffeomorphic via a diffeomorphism fixed at the boundary $\partial \Delta_i$.*

Then the contact half-spaces $(\mathbb{R}^3_+, \xi)$ and $(\mathbb{R}^3_+, \tilde{\xi})$ are isomorphic via a diffeomorphism fixed at the boundary $P = \partial \mathbb{R}^3_+$.

PROOF OF PROPOSITION 3.4.3. Suppose we are given two tight contact structures on $\mathbb{R}^3_+$, the standard one $\xi = \{dz = y dx\}$, and another one $\tilde{\xi}$, so that both structures coincide in a neighborhood U of the boundary $P = \{y = 0\}$. Let X be a contact vector field for the contact structure $\tilde{\xi}$ which is transversal to P and forward integrable. We can also assume that in the neighborhood $U \supset P$ the vector field X coincides with the vector field $X_0 = y \frac{\partial}{\partial y} + z \frac{\partial}{\partial z}$ which is contact for the contact structure ξ. Consider a function $\varphi : \mathbb{R}^2 \to \mathbb{R}_+$, positive on the disk $D = D(\frac{1}{2})$ of radius $\frac{1}{2}$, with the support in the disk $D(1) \subset P$ of radius 1. For $n = 1, \ldots$ let us set

$$\varphi_n(x, z) = n\varphi(\frac{x}{n}, \frac{z}{n}), \ \Omega_n = \{(x, y, z) \mid 0 < y < \varphi_n(x, z)\},$$

$$D_n = \{y = \varphi_n(x, z); \ (x, z) \in D(n)\}.$$

There exists an integer $N = N(n)$ such that it takes time $t > 2$ to reach the domain Ω_n along any trajectory of the vector fields X and X_0 which begins at a point $(x, 0, z) \in P$ with $|x| > N(n)$. In particular, the sets

$$\{X_0^t(x, 0, z)\}_{z < -N(n), \ t \in [0,2]} \quad \text{and} \quad \{X^t(x, 0, z)\}_{z < -N(n), \ t \in [0,2]}$$

contain the parallelpiped

$$I(n) = \{-N(n) - 2n \leq x \leq -N(n), \ 0 \leq y \leq 1, \ -n \leq z \leq n\}.$$

This allows us to find a contact embedding

$$f : (I(n), \xi) \to (\mathbb{R}^3_+ \setminus \bar{\Omega}_n, \tilde{\xi})$$

which is the identity near $P \cap I(n)$. Take a function $\alpha_n : P \to \mathbb{R}_+$ with the support in $P \cap I(n)$ such that the domain $\{0 < y < \alpha_n(x, z)\}$ fills almost all the parallelpiped $I(n)$. Notice that the holonomy map $h_n = h_{\alpha_n} : \mathbb{R} \to \mathbb{R}$, which is defined as at the beginning of this section by the characteristic foliation induced by the contact structure ξ on the surface $\{y = \alpha_n(x, z)\}$, is supported in the interval $\{-n < z < n\}$, and we have

$$h_n(\{-n < z < -n + \varepsilon\}) \supset \{-n < z < n - \varepsilon\},$$

where $\varepsilon > 0$ can be made arbitrarily small by an appropriate choice of the function α_n. Let Σ_n be the graph

$$\{y = \alpha_n(x, z) + \varphi_{n-1}(x, z); \ (x, z) \in P\}.$$

Set

$$\tilde{\Sigma}_n = (\Sigma_n \setminus I(n)) \cup f(\Sigma_n \cap I(n)).$$

This is a smooth plane, properly embedded into $\mathbb{R}^3_+$. Notice that the characteristic foliations $\mathcal{F}_n$, induced by ξ on Σ, and $\tilde{\mathcal{F}}_n$, induced by $\tilde{\xi}$ on $\tilde{\Sigma}$, are diffeomorphic away from D_n. We will keep the notation f for this diffeomorphism, which we set equal to the identity on $P \cap \Sigma_n = P \cap \tilde{\Sigma}_n$.

Next, by a C^0-small perturbation of $P \cap \Sigma_n = P \cap \tilde{\Sigma}_n$ inside the neighborhood $U \subset \mathbb{R}^3_+$ we can create an elliptic-hyperbolic pair (E, H) of singular points of the characteristic foliations $\mathcal{F}_n$ and $\tilde{\mathcal{F}}_n$ (recall that ξ and $\tilde{\xi}$ coincide in the neighborhood U), so that the leaves of both characteristic foliations $\mathcal{F}_n$ and $\tilde{\mathcal{F}}_n$ which intersect the interval $\{x = -N(n), \ -n < z < n - \varepsilon\}$ are being focused at the elliptic point E, see Figure 3.2.

There exists a curve $\Gamma_n \subset \Sigma_n$, transversal to the characteristic foliation $\mathcal{F}_n$, such that the disk Δ_n bounded by this curve contains the elliptic point E and the disk D_{n-1}. The disc $\tilde{\Delta}_n$ bounded by the transversal curve $f(\Gamma_n)$ contains the disk D_{n-1} and the point E as well. Both transversal curves Γ_n and $f(\Gamma_n)$ are transversally isotopic within each own contact structure to the same small transversal curve surrounding the elliptic point E. In particular,

$$e(\xi)[\Delta_n] = e(\tilde{\xi})[\tilde{\Delta}_n] = -1.$$

According to 3.4.4B. we can deform the surface $\tilde{\Delta}_n$ via a C^0-small isotopy in order to kill all the singularities of the characteristic foliation $\tilde{\mathcal{F}}_n|_{\tilde{\Delta}_n}$ except one elliptic point. After this perturbation the characteristic foliations on Σ_n and $\tilde{\Sigma}_n$ become homeomorphic via a homeomorphism, still denoted by f, which is fixed at infinity.

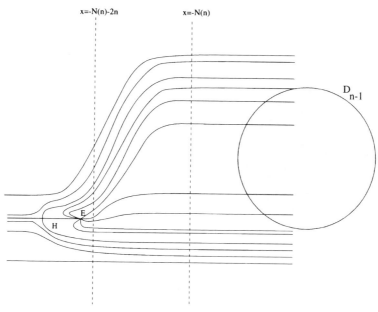

x=-N(n)-2n   x=-N(n)

D_{n-1}

E

H

FIGURE 3.2. Characteristic foliations $\mathcal{F}_n$ and $\tilde{\mathcal{F}}_n$ after creation of the pair (E, H) of singularities. The foliations differ only inside the disk D_{n-1} .

According to 3.4.4A. one can arrange *diffeomorphism* of the characteristic foliations via an additional C^1-perturbation of the surface $\tilde{\Sigma}_n$.

We can define now inductively two sequences of surfaces $S_n, \tilde{S}_n \subset R^3_+$, which are flat at infinity, and such that the characteristic foliations induced on S_n and $\tilde{S}_n$ by ξ and $\tilde{\xi}$ respectively, are diffeomorphic via a compactly supported diffeomorphism f_n. Set $S_1 = \Sigma_1$, $\tilde{S}_2 = \tilde{\Sigma}_1$. Suppose that the surfaces S_i and $\tilde{S}_i$ for $i = 1, \ldots, k-1$, are already chosen. Let K be a compact set in R^3_+ such that outside of K the surfaces S_{k-1} and $\tilde{S}_{k-1}$ are flat, and outside of $K \cap P$ the diffeomorphism f_{k-1} is the identity. For a sufficiently large $n = n(k)$ we have $\Sigma_n \cap K = \tilde{\Sigma} \cap K = \emptyset$. Thus we can define $S_k = \Sigma_{n(k)}$, $\tilde{S}_k = \tilde{\Sigma}_{n(k)}$. To finish the proof it remains to apply Lemma 3.4.5.   □

3.5. Tight and Taut confoliations

In this section we will discuss various generalizations of tightness and tautness to confoliations. These conditions sound reasonable and interesting to us, but this area is not yet mature — it is in need of further constructions to give it more body and further theorems to bind it together, so it is not unlikely that the best definitions are yet to be found, A number of lines of further investigation are suggested by material in this section, and some of the proofs we will give are rather sketchy, but we hope that this discussion of mathematics that is still in an immature state will help stimulate further work.

We will define below the notions of tight and taut confoliations. Conjecturally, our taut confoliations form a subclass of tight confoliations. We do this with some reluctance, since we are afraid that using both words, tight and taut, with distinct meanings but in a very close context can create confusion. It seems that people can be divided into two camps regarding the use of these terms. Some think that tight is tighter than taut, while the others believe that taut is tauter. It could be that both camps are right.

3.5.1. Tight. The following notion interpolates between the tight contact structures and foliations without Reeb components.

DEFINITION 3.5.1. A confoliation (M, ξ) is tight if for every embedded 2-disk $D \subset M$ satisfying

- ∂D is tangent to ξ and
- D itself is transverse to ξ in a neighborhood of ∂D

there is another disk D' such that

- $\partial D' = \partial D$,
- D' is tangent to ξ, and
- $e(\xi)[D \cup D'] = 0$.

Of course, in the case of a contact structure a disk D' tangent to ξ cannot exist, hence the above tightness condition for the contact case coincides with the definition of a tight contact structure from Section 3.1.

For a foliation, this definition of tightness just means the absence of vanishing cycles, which for closed manifolds is equivalent to the absence of Reeb components, that is, a *Reebless* foliation. A tight or Reebless foliation is not necessarily taut (see Section 3.1 above), but in this case it contains closed torus leaves which separate the manifold into taut pieces.

Theorem 3.2.4 admits the following generalization

THEOREM 3.5.1. *Any symplectically semi-fillable confoliation on a 3-manifold is tight.*

Note that the manifold is not assumed to be compact.

PROOF. Similarly to the proof of Theorem 3.2.4, the proof of this result uses the method of filling with holomorphic disks. Choose an almost complex structure J in W tamed by ω (i.e. the quadratic form $\omega(X, JX)$ is positive definite, see [**30**]), and such that ξ is J-invariant. This is possible in view of the condition that ω dominates ξ. In fact, according to Proposition 1.1.9 one can even make J integrable near the boundary. Then the orientation condition shows that with such a choice of the orientation the boundary $M = \partial W$ is weakly J-convex. Without loss of generality (see [**43**]) we can assume that (W, ω) is minimal and, in particular, (W, J) contains no J-holomorphic spheres.

Suppose there exists a disk $D \subset M$ such that $\Gamma = \partial D$ is tangent to ξ but D is transverse to ξ along ∂D. Thickening the disk, we can find a 2-sphere $S \subset M$ which contains the disk D. The characteristic foliation S_ξ has Γ as a closed leaf.

We want to apply the technique of filling with holomorphic disks (see [**30**], [**13**] and [**60**]). However, the weak J-convexity condition, instead of strict pseudocon-vexity, which was used in [**30**], [**13**] and [**60**] brings additional analytical difficulties. The following result is proven by R. Hind (see [**34**]).

THEOREM 3.5.2. *Let W, M, ξ, J, ω be as above and S, $S \subset M$, be an embedded 2-sphere. Then, maybe after a C^2-perturbation in a neighborhood of its complex points, the sphere S can be filled with holomorphic disks in the following sense. There exists a 1-dimensional foliation $\mathcal{H}$ of S with singularities in complex points of the sphere S. Non-singular leaves of $\mathcal{H}$ are smooth circles. Singular leaves are either (elliptic) points or homoclinic orbits of hyperbolic points. Each non-singular, or homoclinic leaf of $\mathcal{H}$ bounds a holomorphic disk, embedded into W. Each leaf of $\mathcal{H}$ is either transversal to the leaves of the characteristic foliation ξ, or coincides with one of these leaves. In the latter case, the holomorphic disk bounded by this leaf is contained in the boundary $M = \partial W$.*

REMARK 3.5.3. As in the strictly J-convex case the holomorphic disks which are not contained in the boundary are disjoint. However, those in the boundary may contain one another.

Theorem 3.5.1 follows now from Proposition 3.5.2. Let us apply 3.5.2 to fill the sphere S with holomorphic disks. The closed leaf Γ of the foliation S_ξ cannot be transversal to the foliation $\mathcal{H}$. Hence it coincides with one of the leaves of the characteristic foliation $\mathcal{H}$ and, therefore bounds a holomorphic disk D' inside M, which is, necessarily, an integral disk of ξ. The union $D \cup D'$ is homological to 0 inside W. Hence,

$$e(\xi)[D \cup D'] = c_1(W)[D \cup D'] = 0,$$

where $c_1(W)$ is the first Chern class of the almost complex manifold (W, J). ☐

It seems likely that combining the methods from [**53**], [**56**] and [**12**] one can prove the following generalization of Theorem 3.3.1.

CONJECTURE 3.5.4. *The inequalities (3.1) and (3.3) hold for tight confoliations, i.e tight confoliations belong to the class $\mathcal{T}$ defined in the previous section.*

In Section 3.5.4 below we show that a perturbation of a tight foliation is *strongly tight*, and, in particular, belongs to the class $\mathcal{T}$.

3.5.2. Tightness and coverings. Let $f : (M_1, \xi_1) \to (M_2, \xi_2)$ be a covering map between two confoliated manifolds, such that the confoliation ξ_2 is the push-forward of the confoliation ξ_1 under the projection f. Notice that *if ξ_1 is tight (taut), then ξ_2 is also tight (taut)*.

A finite cover of a taut foliation is taut, but a cover a tight contact structure need not to be tight even in the case of finitely-sheeted cover. The first example of this kind was constructed by S. Makar-Limanov in [**41**]. Several other examples were constructed by R. Gompf, see [**27**]. On the other hand tightness of contact structures which can be obtained by a perturbation of taut foliations survive the passage to the universal covering (see 3.2.8 above). Therefore, it seems sensible to distinguish the subclass of tight contact structures, and more generally tight confoliations, which survive the passage to the universal cover. Hence we define

DEFINITION 3.5.2. A confoliation ξ on 3-manifold M^3 is *strongly tight* when

- the universal covering space of M^3 is $\mathbb{R}^3$, and
- the confoliation induced on the universal cover is tight.

In the case of foliations, the only difference between tight and strongly tight is that we have forbidden $S^2 \times S^1$. A s it was already pointed out in the previous section, any tight contact structure on $\mathbb{R}^3$ is isotopic to the standard contact structure (see [16]). [2]

According to 3.2.7 universal covers of taut foliations are symplectically semifillable, and hence tight. On the other hand, manifolds carrying taut foliations (and, moreover, Reebless foliations) are covered by $\mathbb{R}^3$ (see [49]). Thus we have

PROPOSITION 3.5.5. *A confoliation, C^0-close to a taut foliation, is strongly tight.*

3.5.3. Criterion for strong tightness. In order to identify strongly tight confoliations in general, we need criteria to recognize tight confoliations on $\mathbb{R}^3$.

Suppose a confoliation ξ is transverse to fibers of the standard fibration $\mathbb{R}^3 \to \mathbb{R}^2$ consisting of projection in the z-direction. Then ξ can be thought of as a connection for this $\mathbb{R}$-bundle, although parallel translation *a priori* is only locally defined. We will call ξ a *complete connection* if parallel translation is globally defined, *i.e.*, for any smooth (or rectifiable) path p in $\mathbb{R}^2$ and any lift of its initial point to $\mathbb{R}^3$, there is an extension to a lift of p tangent to ξ.

PROPOSITION 3.5.6. *If a confoliation ξ in $\mathbb{R}^3$ is a complete connection for the fibration $\mathbb{R}^3 \to \mathbb{R}^2$, then it is tight in the sense of section 3.5.1 above.*

PROOF. In the case of a foliation or a contact structure, this follows quickly because the condition that ξ be a complete connection implies that ξ is standard. For the general confoliations let us first observe that the complete connection condition implies that the leaves of the characteristic foliation on a plane $\{x = \text{const}\}$ are graphs of functions $z = h(y)$, $y \in (-\infty, \infty)$. Thus changing the confoliation $\xi|_C$ by a diffeomorphism preserving the fibers of the fibration $\mathbb{R}^3 \to \mathbb{R}^2$ we can make all the leaves horizontal, i.e. parallel to the y-axis. This means that the deformed confoliation, still denoted by ξ, can be given by an equation $dz = f(x,y)dx$, and hence the confoliation condition just says that $\frac{\partial f}{\partial y}(x,y) \geq 0$, see Criterion 1.1.5. It is clear from the "propeller" description of a confoliation that for any compact set $K \subset \mathbb{R}^3$ there exists a confoliation $\tilde{\xi} = \{dz = \tilde{f}(x,y) = 0\}$, which coincides with ξ on K, and which coincides with the standard contact structure (i.e. $\tilde{f}(x,y) = y$) at infinity. But the standard contact structure on $\mathbb{R}^3$ is isomorphic to the standard contact structure on $S^3 \setminus \{\text{point}\}$, defined by complex tangencies to the unit sphere in $\mathbb{C}^2$. Hence we can implant ξ in a ball in S^3, so that the new confoliation $\tilde{\xi}$ is transversal to the fibers of the Hopf fibration. That means that the standard symplectic form in $\mathbb{C}^2$ dominates $\tilde{\xi}$, that is $\tilde{\xi}$ is symplectically fillable, and hence tight. Hence, $\xi|_K$ is tight, and because the compact set K is arbitrary, the confoliation ξ is tight as well.   □

REMARK 3.5.7. Without the completeness condition, 3.5.6 is false. For example, if you turbularize the foliation of T^3 by horizontal 2-tori along a vertical curve,

---

[2]Modulo the geometrization conjecture, universal coverings of prime 3-manifolds are diffeomorphic to either $\mathbb{R}^3$, or $S^2 \times \mathbb{R}$ or S^3. All these three manifolds admit only one tight contact structure up to isotopy (see [12], [16]). We have singled out the case $\mathbb{R}^3$ in the definition of strong tightness because this is the generic case in 3-manifold topology, and other cases tend to be rather special. Thus, one should keep in mind that the other two interesting classes of tight contact structures are quotients of the standard contact $S^2 \times \mathbb{R}$ and S^3.

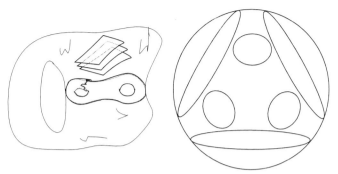

FIGURE 3.3. Branching of foliations in universal cover. A foliation
of a closed 3-manifold M with one closed leaf S is schematically
represented on the left. The universal cover $\tilde{M}$ is represented on
the right; assuming S is not the fiber of a fibration, the copies of the
universal cover of $\tilde{S}$ separate $\tilde{M}$ into countably many components,
each having boundary a countable union of copies of $\tilde{S}$.

the vertical flow can be perturbed to give a flow on T^3 transverse to the resulting foliation whose lift to the universal cover $\mathbb{R}^3$ is still predominantly vertical, so
equivalent to the to the standard fibration. A similar construction works to show
that there are overtwisted contact structures in $\mathbb{R}^3$ transverse to the vertical lines.

The complete connection condition is sufficient, but in the case of foliations is
not necessary, for tightness in $\mathbb{R}^3$. The leaves of the universal cover of a strongly
tight foliation ξ are properly embedded planes, but this does not imply that the
foliation is necessarily equivalent to a foliation transverse to the vertical line field.
Indeed, the space of leaves in the universal cover even of a taut foliation can branch.
For example, if ξ has a closed leaf S which is not the fiber of some fibration over
S^1, then the preimage of S in the universal cover $\mathbb{R}^3$ necessarily separates $\mathbb{R}^3$ in a
non-trivial tree pattern (Figure 3.3).

It is curious that this complicated branching pattern, after perturbation to be
a tight contact structure, becomes equivalent to the standard contact structure.

Let ξ be a confoliation of a manifold M whose contact part is saturated; this
seems a reasonable stipulation in view of Section 2.8. To decide if ξ is strongly tight,
we can look separately at the contact part and the lamination $C(\xi)$ which is the
fully-foliated part. For ξ to be strongly tight, each component of the contact part
must have universal cover equivalent to the standard contact structure on $\mathbb{R}^3$. The
lamination $C(\xi)$ must be an essential lamination, which implies its universal cover
has leaves which are properly embedded planes that separate the universal cover in
a tree-like or dendritic pattern. Conversely, if $C(\xi)$ is an essential lamination and
if each contact part has standard universal cover, then the confoliation is strongly
tight, since if the boundary of a disk is tangent to the confoliation of $\mathbb{R}^3$, then the
boundary must either lie on a leaf or in one of the contact components. The leaves
of an essential lamination are planes, so any such disk can be homotoped either
onto a leaf or into a contact component.

The theory of sutured manifolds and essential laminations gives the following,
using techniques developed by David Gabai:

PROPOSITION 3.5.8. *Let ξ be a tight confoliation of an irreducible 3-manifold M^3. If ξ is not transitive, that is, if the fully-foliated part $C(\xi)$ is non-empty, then there is a strongly tight foliation ξ' (which may only of class C^0) coinciding with ξ on its fully-foliated part.*

REMARK 3.5.9. Note that the proposition makes no claim as to the relation of ξ with ξ' except that they agree on $C(\xi)$. It does not answer the interesting question of what circumstances are necessary for ξ can be deformed to a taut foliation through tight contact structures.

To prove this proposition, one finds an essential branched surface that approximates the fully-foliated lamination of ξ. The complement of the branched surface is a 3-manifold with boundary having a seam or *suture* where two branches of the branched surface join together. It is convenient to represent this by thickening the seam into a cylinder. Each non-suture portion of the boundary has minimum genus in its homology class, by the inequalities (3.1) and (3.3). Gabai proved that under those circumstances, one can construct a foliation of the sutured manifold where the non-suture portion of its boundary consists of leaves and the cylindrical sutures are transverse, with the induced foliation being a product. These foliations extend to foliate the complement of the fully-foliated lamination for ξ, creating a foliation ξ'.   □

3.5.4. Gluing tight contact manifolds.

Suppose a surface F divides a contact 3-manifold M into two (strongly) tight [3] parts: $M = M_1 \cup M_2$, $M_1 \cap M_2 = F$. Is M tight? Here F can be a closed, or properly embedded non-compact surface.

The answer is positive for gluing tight contact structures along 2-spheres. This was independently observed by V. Colin ([9]) and S. Makar-Limanov ([42]).

In the general case the answer is negative: for instance, by adding a tight torical annulus $T^2 \times I$ to a tight solid torus, it is easy to get an overtwisted result. However, Theorem 3.5.11below , which was first proven by V. Colin [10], shows that the strong tightness is preserved when one glues along incompressible tori.

We begin with a corollary of Proposition 3.4.3.

COROLLARY 3.5.10. *Suppose that a convex simple plane P divides a contact manifold M into two strongly tight parts: $M = M_1 \cup M_2$, $M_1 \cap M_2 = P$. Then M is strongly tight.*

PROOF. Without loss of generality we can assume that M_1 and M_2 are simply-connected. Then they can be realized as domains in the standard contact $\mathbb{R}^3$. The copies P' and P'' of the simple plane P serve as boundary components of these domains. Let $\widehat{M_1}$ and $\widehat{M_2}$ be half-spaces in $\mathbb{R}^3$ bounded by P' and P'' and such that $\widehat{M_i} \supset M_i$ for $i = 1, 2$. These half-spaces are bounded by semi-convex planes (for the co-orientation given by interior transversals to the boundary. Hence, in view of Proposition 3.4.3 both $\widehat{M_1}$ and $\widehat{M_2}$ are contactomorphic to the standard half-space, and therefore we can identify $\widehat{M_1}$ with the upper half-space $\{y \geq 0\}$, and $\widehat{M_2}$ with the lower half-space $\{y \leq 0\}$. This identification allows us to realize $M = M_1 \cup M_2$ as a domain in the standard contact space $\mathbb{R}^3$, and therefore M is (strongly) tight.   □

---

[3]We call a manifold with boundary tight if it extends to a larger tight contact manifold without boundary.

THEOREM 3.5.11. (V. Colin, [10]) *Let T be an incompressible torus in a contact manifold (M, ξ). Let us cut open M along the torus T. If the contact structure ξ on the resulted manifold $\widetilde{M}$ with boundary is strongly tight, then ξ is strongly tight.*

PROOF. According to 3.4.4B. we can perturb the torus T by a C^0-small isotopy to obtain a torus T' transversal to the contact structure ξ and such that the characteristic foliation induced by ξ on the torus T' admits a global transversal circle S, so that the characteristic foliation is the suspension of a diffeomorphism $g : S \to S$. By a C^∞-small additional perturbation we can arrange that the diffeomorphism g has rational rotation number, and moreover, that it has isolated nondegenerate periodic points. According to Giroux (see [24]), this property garantees *contact convexity*, i.e. existence of a contact vector field X, transversal to T'. One can extend X to the whole M cutting it to 0 outside an arbitrarily small compact neighborhood of T'. Thus the field X is complete. Clearly, the manifold M_1 with boundary which one obtains by cutting open M along T' is still strongly tight. According to the definition of strongly tight contact structure the universal cover $\widehat{M}_1$ of M_1 can be realized as a domain in the standard contact $\mathbb{R}^3$ bounded by planes $P_1, P_2, \ldots$, covering the torus T'. The structure of the characteristic foliation on T' ensures that all these planes are simple. Moreover, these planes are semi-convex, being co-oriented by internal normal fields. Indeed, the required complete contact vector field transversal to P_i can be constructed by lifting the contact vector field X. Thus according to Proposition 3.4.3 for each $i = 1, \ldots,$ the half-space E_i bounded by P_i, which contains $\widehat{M}_1$ is contactomorphic to the standard half-space $\mathbb{R}^3_+$. The universal cover $\widehat{M}$ of M can be built by gluing several copies of $\widehat{M}_1$ along planes P_i in a tree-like manner, i.e. any two copies are glued along at most one of their boundary planes. Thus it is sufficient to show that one get a strongly tight contact manifold by gluing two copies $\widehat{M}_1$ and $\widehat{M}'_1$ along their boundary components $P \subset \widehat{M}_1$ and $P' \subset \widehat{M}'_1$, which are simple planes. Hence strong tightness of $\widehat{M}$ follows from 3.5.10. □

COROLLARY 3.5.12. *A contact structure ξ, which is C^0-close to a tight (= Reebless) foliation ξ_0 is strongly tight.*

PROOF. We can assume that ξ_0 has finitely many torus leaves. For each torus leaf T_i of ξ_0, we can find a regular neighborhood whose boundary is transverse to ξ_0. The foliation ξ_0 is taut on the complement $M \setminus \bigcup N_i$. Thus according to Corollary 3.5.5 the contact structure $\xi|_{M \setminus \bigcup_i N_i}$ is strongly tight, provided that it is sufficiently close to ξ_0. On the other hand the covering of the contact structure $\xi|_{N_i}$ is a complete connection on $\mathbb{R}^3$, and hence it is tight according to 3.5.6. Therefore $\xi|_{N_i}$ is strongly tight, and hence Theorem 3.5.11 implies that ξ is strongly tight as well. □

3.5.5. Taut. H. Hofer proved in [35] the following criterion of tightness of contact structures.

THEOREM 3.5.13. (H. Hofer, [35]) *Suppose that a contact structure ξ admits a contact form α such that its Reeb vector field (i.e. the vector directing the kernel of the 2-form $d\alpha$) has no contractible periodic orbits. Then ξ is tight.*

In fact, similar arguments prove more: under the assumptions of 3.5.13 ξ is strongly tight. It sounds likely that Hofer's theorem can be generalized as follows.

CONJECTURE 3.5.14. *Suppose a confoliation structure ξ on a closed manifold is dominated by a symplectic form ω such that the vector field X which generates $\operatorname{Ker}\omega$ has no contractible periodic orbits. Then ξ is strongly tight.*

Note that, from remark 3.5.7, on non-compact manifolds (e.g. $\mathbb{R}^3$) there are confoliations which are not tight that have transverse Hamiltonian vector fields with no periodic orbits.

As mentioned in Section 3.2, taut foliations are exactly those which admit a dominating symplectic form; the absence of contractible periodic orbits is automatic in this case. This motivates the following definition:

DEFINITION 3.5.3. *A confoliation (on a compact manifold) is taut if it is dominated by a symplectic form ω such that the vector field X which generates $\operatorname{Ker}\omega$ has no contractible closed orbits.*

In the case of foliations, the definition of taut coincides with with the previous definition of tautness, and is strictly stronger than strongly tight. However, for contact structures this implication in general case is unknown, and it is the subject of the above conjecture 3.5.14.

3.6. Homotopy of confoliations

THEOREM 3.6.1. *Classification of confoliations up to homotopy in the class of confoliations coincides with their homotopical classification as tangent plane fields.*

Let us begin with the following

PROPOSITION 3.6.2. *Any confoliation is homotopic to an overtwisted contact structure in the class of confoliations.*

PROOF. Let Γ be a closed curve transversal to a confoliation ξ. One can choose a splitting $U = S^1 \times D^2$ of a tubular neighborhood $U \supset \Gamma$ such that in coordinates $\varphi \in S^1$, $(\rho, \theta) \in D^2$ the confoliation ξ can be defined by the 1-form

$$\alpha = d\varphi + f(\rho, \theta, \varphi)d\theta,$$

where

$$\frac{\partial f}{\partial \rho} \geq 0.$$

Let $\delta : I \to I$ be a C^∞ function which is equal to 0 on $[0, 1/3]$, equals 1 on $[2/3, 1]$ and has positive derivative on $(1/3, 2/3)$. Then the family of forms

$$\alpha_t = t\alpha + (1-t)(d\varphi + \delta(\rho)f(\rho, \theta, \varphi)d\theta)$$

defines on U a deformation in the class of confoliations of $\xi|_U$, fixed near ∂U into a confoliation ξ_0 whose restriction to a smaller neighborhood $U' \supset \Gamma$ is a fibration by normal disks.

The next step is to deform the foliation $\xi_0|_{U'}$, as it is shown on Figure 3.4 into a foliation ξ' which has a Reeb component R along the curve Γ and a foliated I-bundle $B = T \times I$ attached to R along the torus $T = \partial R$.

The foliation $\xi'|_B$ has non-trivial holonomy along the meridian $\Delta \subset T$. Choosing this holonomy dilating or attracting we can arrange that a slightly extended

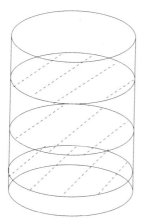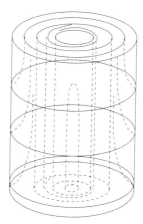

FIGURE 3.4. Deformation of the foliation $\xi_0|_{U'}$.

meridianal disk D of the solid torus R is bounded by a transversal curve which vi-
olates the inequality (3.3). This guarantees that any contact structure sufficiently
close to ξ' should be overtwisted. Let us do this kind of deformation along sufficiently many closed transversals, to destroy all closed leaves except couples of tori
created by our deformation along each of the chosen transversals. On each of the
created torus leaves there exists, by construction, a curve with non-trivial attracting holonomy. Inserting, if necessary, an appropriate foliated I-bundle instead of
each closed leaf we can arrange that each closed leaf of the deformed foliation has
a curve with attracting holonomy on both sides. Moreover, this holonomy can be
arranged to be nice (see Section 2.9). Thus we can apply Proposition 2.9.2 to deform this confoliation into a contact structure ξ'' which is overtwisted, as it was
indicated above.   □

Theorem 3.6.1 is an immediate corollary of 3.6.2 and the classification of overtwisted contact structures from [11].

Theorem 3.6.1 shows that classification of confoliations up to homotopy is not
very interesting. However, an important question remains about the classification
of tight confoliations up to a homotopy through tight confoliations. See Problem 3
in the next section.

3.7. A few open problems about confoliations

1. Is tightness an open condition for confoliations (in C^k-topology for $k = 0, \ldots, \infty$)?
2. Example 3.3.6 shows that tightness is not a closed condition among the
   confoliations. Describe the class of foliations which are limits of tight contact
   structures.
3. Classify tight confoliations up to homotopy in this class. Is there always
   a finite number of homotopy classes of tight confoliations? Notice that for
   tight contact structures the answer is negative.
4. Is it possible to C^k-deform a C^k-foliation, $k = 0, 1, \ldots, \infty$, into a contact
   structure?

5. Suppose a positive contact structure ξ admits a transversal negative contact structure. Is ξ tight?

6. Any overtwisted contact structure ξ arises as a deformation of a foliation, *i.e.* there exists a family of confoliations ξ_t, $t \in \mathbb{R}_+$, such that ξ_0 is a foliation and ξ_t for $t > 0$ is a contact structure isotopic to ξ. What is the situation for tight contact structures?

7. Which orientable 3-manifolds admit tight confoliations (or contact structures)? At the moment no restrictions are known.

8. Are there tight confoliations which are not symplectically (semi-) fillable?

Bibliography

[1] V. I. Arnold, private communication, 1996.

[2] V.I. Arnold and A.B. Givental, Symplectic geometry, in the book *Dynamical systems*, vol. 4, Springer-Verlag, 1988.

[3] S. Altschuler, A geometric heat flow for one-forms on three-dimensional manifolds, *Illinois J. of Math.* **39**(1995), 98–118.

[4] D.V. Anosov, Geodesic flows on closed Riemannian manifolds with negative curvature, *Trudy MIAN*, **90**(1969).

[5] A. Denjoy, Sur les courbes définis par les équations différentielles sur la surface du tore, *J. de Math.*, **11**(1932), 333–375.

[6] C. Bonatti and S. Firmo, Feuilles compactes d'un feuilletés genériques en codimension 1, *Ann. Sci. Ecole Norm. Sup.*, **27**(1994), 407–462.

[7] D. Bennequin, Entrelacements et équations de Pfaff, *Astérisque*, **107-108**(1983), 83–161.

[8] A. Candel, Uniformization of surface laminations, *Ann. Sci. École Norm. Sup.*, 4, **26**(1993), 489–516.

[9] V. Colin, Chirurgies d'indice un et isotopies de sphéres, preprint 1996.

[10] V. Colin, Recollement des variétés de contact tendues, in preparation.

[11] Y. Eliashberg, Classification of overtwisted contact structures, *Invent. Math.*, **98**(1989), 623–637.

[12] Y. Eliashberg, Contact 3-manifolds twenty years after J. Martinet's work, *Annales de l'Inst. Fourier*, **42**(1992), 165–192.

[13] Y. Eliashberg, Filling by holomorphic discs and its applications, *Lect. Notes LMS*, **151**(1992), 45–67.

[14] Y. Eliashberg, Legendrian and transversal knots in tigh contact manifolds, in the book *Topological methods in modern Mathematics*, Publish or Perish, 1993.

[15] Y. Eliashberg, Topological characterization of Stein manifolds of complex dimension > 2, *Int. J. of Math*, **1**(1991), 29–46.

[16] Y. Eliashberg, Classification of contact structures on $\mathbb{R}^3$, *Int. Math. Res. Notices*, **3**(1993), 87–91.

[17] Y. Eliashberg and M. Gromov, Convex symplectic manifolds, *Proc. Symp. Pure Math.*, **52**(1991), Part 2, 135–162.

[18] Y. Eliashberg and L. Polterovich, New applications of Luttinger's surgery, Comm. Math. Helvet., **69**(1974), 512–522.

[19] J. Franks and R. Williams, Anomalous Anosov flows, *Lecture notes in Math.*, Springer, **819**(1980), 158–174.

[20] D. Gabai, Foliations and the topology of 3-manifolds, *J. Diff. Geometry*, **18**(1983), 445–503.

[21] D. Gabai and U. Oertel, Essential laminations in 3-manifolds, *Annals of Math.*, **130**(1989), 41–73.

[22] H. Geiges, Examples of symplectic 4-manifolds with disconnected boundary of contact type, *Bull. London Math. Soc.*, **27**(1995), 278–280.

[23] E. Giroux, Convexité en topologie de contact, *Comm. Math. Helvet.*, **66**(1991), 637-677.

[24] E. Giroux, Une structure de contact, même tendue est plus ou moins tordue, *Ann. Scient. Ec. Norm. Sup.*, **27**(1994), 697–705.

[25] E. Giroux, Structures de contact sur les fibrés aux circles, preprint 1994.

[26] E. Ghys, private communication.

[27] R. Gompf, Handlebody constructions of Stein surfaces, preprint 1996, to appear in *Annals of Math.*.

[28] S. Goodman, Closed leaves in foliations of codimension one, *Comm. Math. Helv*, **50**(1975), 383–388.

[29] J.W. Gray, Some global properties of contact structures, *Annals of Math.*, **69**(1959), 421–450.

[30] M. Gromov, Pseudo-holomorphic curves in symplectic manifolds, *Invent. Math.*, **82**(1985), 307–347.

[31] R. Gunning and H. Rossi, Analytic functions of several complex variables, Prentice-Hall, 1965.

[32] A. Hatcher, Some examples of essential laminations in 3-manifolds, *Ann. Inst. Fourier*, **42**(1992), 313–325.

[33] G. Hector and U. Hirsch, Introduction to the geometry of foliations, Parts A and B, Vieweg, 1983.

[34] R. Hind, Filling by holomorphic discs with weakly pseudo-convex boundary conditions, to appear in *GAFA*.

[35] H. Hofer, Pseudo-holomorphic curves and Weinstein conjecture in dimension three, *Invent. Math.*, **114**(1993), 515–563.

[36] H. Jacobowitz, An introduction to CR-structures, AMS, Providence, 1990.

[37] Y. Kanda, The classification of tight contact structures on the 3-torus, preprint 1995.

[38] N. Kopell, Commuting diffeomorphisms, *Proc. Sympos. Pure Math.*, **14**(1968), 165–184.

[39] P. Kronheimer and T. Mrowka, Monopoles and contact structures, preprint 1996, to appear in *Invent. Math.*

[40] R. Lutz, Sur la géometrie des structures de contact invariantes, *Ann. Inst. Fourier*, **29**(1979), 283–306.

[41] S. Makar-Limanov, Tight contact structures on 3-dimensional solid tori, to appear in *Trans. AMS.*

[42] S. Makar-Limanov, Morse surgeries of index 0 and 1 on tight manifolds, preprint 1996.

[43] D. McDuff, Structure of rational and ruled symplectic 4-manifolds, *J. of AMS*, **3**1990, 679–712.

[44] J. Milnor, On the existence of a connection with curvature 0, *Comm. Math. Helvet.*, **32**(1958), 215–223.

[45] K. Mishachev, Lagrangian and Legendrian fibrations, Master Thesis, Moscow State University, 1996.

[46] Y. Mitsumatsu, Anosov flows and non-Stein symplectic manifolds, *Ann. Inst. Fourier*, **45**(1995), 1407–1421.

[47] S. Morita and T. Tsuboi, The Godbillon-Vey class of codimension one foliations without holonomy, *Topology*, **19**(1980), 43–49.

[48] S. P. Novikov, Topology of foliations, *Trans. Moscow Math. Soc.*, **14**(1963), 268–305.

[49] C. Palmeira, Open manifolds foliated by planes, *Ann. of Math.*, **107**(1978), 109–131.

[50] A. Sato and T. Tsuboi, Contact structures of closed manifolds fibered by the circles, *Mem. Inst. sci. Tech. Meiji Univ.* **33**(1994), 41–46.

[51] R. Sacksteder, Foliations and pseudogroups, *Amer. J. of Math.*, **87**(1965), 79–102.

[52] D. Sullivan, A homological characterization of foliations consisting of minimal surfaces, *Comment. Math. Helv.*, **54**(1979), 218–223.

[53] W.P. Thurston, Norm on the homology of 3-manifolds, *Memoirs of the AMS*, **339**(1986), 99–130.

[54] W.P. Thurston, Existence of codimension one foliations, *Ann. Math*, **104**(1976), 249–268.

[55] W.P. Thurston, The theory of foliations in codimension greater then one, *Comm. Math. Helvet.*, **49**(1974), 214–231.

[56] W.P. Thurston, Foliations of 3-manifolds which are circle bundles, PhD Thesis, UC Berkeley, 1972.

[57] W.P. Thurston, A generalization of the Reeb stability theorem, *Topology*, **13**(1974), 347–352.

[58] T. Tsuboi, Γ_1-structures avec une seulle feuille, *Astérisque*, **116**(1984), 222–234.

[59] J. Wood, Bundles with totally disconnected structure group, *Comm. Math. Helvet.*, **45**(1971), 357–273.

[60] R. Ye, Filling by holomorphic discs in symplectic 4-manifolds, preprint, 1994, to appear in *Trans of AMS.*